Springer
Tokyo
Berlin
Heidelberg
New York
Barcelona
Hong Kong
London
Milan
Paris
Singapore

Japan Environmental Council (Ed.)

The State of the Environment in Asia

1999/2000

 Springer

Japan Environmental Council (JEC)

AWAJI Takehisa (Editor-in-Chief)
Professor
Faculty of Law
Rikkyo University
3-34-1 Nishiikebukuro, Toshima-ku, Tokyo 171-8501, Japan

TERANISHI Shun'ichi (Editor-in-Chief)
Professor
Graduate School of Economics
Hitotsubashi University
2-1 Naka, Kunitachi, Tokyo 186-8601, Japan

Rick Davis (Translator)
Ashigawa, Japan

ISBN 4-431-70267-9 (Hard Cover) Springer-Verlag Tokyo Berlin Heidelberg New York
ISBN 4-431-70268-7 (Soft Cover)   Springer-Verlag Tokyo Berlin Heidelberg New York

Printed on acid-free paper

Typesetting: Camera-ready by Springer-Verlag Tokyo
Printing and binding: Obun, Japan
SPIN: 10741933 (Hard Cover) 10741941 (Soft Cover)

Preface
Why Asia?

Significance and Purpose of *The State of the Environment in Asia*

1. Asia and the Future of the Global Environment

Western social scientists used to regard Asia as a "stagnant society," and, except for Japan, people have thought of postwar Asia as a region that is typical of the continuing "vicious circle of poverty."[1]

But in contrast to other developing regions, since the latter half of the 1960s or 1970s South Korea, Taiwan, Hong Kong, and Singapore were the first to follow in Japan's footsteps as the Asian NIEs (newly industrializing economies) and joined the ranks of the industrialized countries and regions. Then in the latter half of the 1980s ASEAN nations one by one embarked on the road to industrialization, after which China, India, Vietnam, and other countries began liberalization, so that the entire Asian region became known as the world center of economic growth. Furthermore, Asia's population easily accounts for half the world's, and estimates say it will rise to two-thirds in the first half of the next century.

Asia's real economic growth rate in recent years has been two to three times the world average, and considering the still-increasing trend toward population growth, it is not an overstatement to say that resource and environmental problems, and the future of our global environment, are dependent on what happens in Asia from now on. We believe Asia is a vital issue that cannot be discounted because in recent years rapid modernization is taking this region down the road to industrialization and urbanization that will create environmental problems worse than those in the West and Japan, and transform Asian countries into socioeconomic systems that squander resources and damage their environments. Thus, a reexamination of Asia will provide a window on the future of our global environment.

---

[1]   Gunnar Myrdal, *Asian Drama: An Inquiry into the Poverty of Nations*, An Abridgement by Seth S. King of the Twentieth Century Fund Study, New York, Pantheon, 1971.

2. From Swift Progress to Big Problems

Because in recent years Asia has been called the center of world economic growth, until now global attention focused on the region has regarded it solely as one of swift progress.

But while this East Asia-centered rapid economic growth gave Asia its momentum, industrialization and urbanization were causing environmental and urban problems throughout the region, as well as destroying rural communities and those countries' historical and cultural traditions. Some of the so-called global environmental problems have also grown serious in Asia, such as soil depletion, desertification, rapid tropical deforestation, and widening damage from acid rain. Further, already in 1991 the Asia-Pacific region accounted for 25% of world emissions of the greenhouse gas CO_2, with 11.2% from China, 4.8% from Japan, and 3.1% from India. If nothing is done, estimates say this region will account for 36% in 2025 and over 50% at the end of the 21st century.

Asia now faces what might be called a triad of environmental problems — regional, international, and global — and the big question now is how it can create a framework suited to the region at a time when the world prepares to enter the new millennium. Indeed, the picture of Asia is more and more that of a region faced with a plethora of difficult challenges. And therein is one more reason why our interdisciplinary group of academics felt we had to compile this report to elucidate the situation in Asia.

3. Cooperation and Solidarity among Asian Environmental NGOs

Meanwhile, international organizations including the UN Economic and Social Commission for Asia and the Pacific (ESCAP), the World Bank, and the OECD have begun to recognize the crucial nature of Asia's environmental problems, and they have already issued a number of reports.[2] These reports provide valuable data and information that are commensurate with the issuing organizations' plentiful funding and staffing, and this book has benefited from the large amount of useful information in these and other reports and research preceding our effort.

Yet there are so many things we still do not understand about Asia's environmental problems. Our primary reason for going through with the founding and publication of this NGO-perspective *The State of the Environment in Asia* is to describe Asia's environmental problems, and use this as the basis for elucidating the challenges of environmental policy in this region. We want to show the reality of pollution and environmental problems — the variety of harm and damage, and the many people who suffer as a result, but this reality is almost impossible to find among existing government documents and statistics. Additionally, there is but a paltry store of data relating to environmental problems and available as statistical values, which shows that the vast majority of Asia's environmental and pollution problems have yet to advance beyond the description stage to the analysis stage. Another factor that one cannot ignore is that political circumstances in Asian countries make it difficult to even study the situation. For example, not only do adminis-

trative offices have insufficient data and information, hardly any of this is available to the public.

In view of this situation, the only way to assess environmental problems in Asia for the time being is for independent environmental NGOs, the researchers and experts affiliated with them, and citizens in Asian countries and regions to build networks based on interpersonal relationships of trust, and use these networks to work away at information gathering and analysis. Needless to say, this involves considerable difficulties.

Fortunately, however, the process of compiling this book afforded us with important ideas on how to build networks for mutual cooperation among environmental NGOs, and affiliated researchers and experts in Asia. It was also possible to use this in conducting a number of field studies and workshops, which allowed us to gather valuable information. Much of this, however, is only fragmentary, making it far too early to paint the big picture.

This 1999/2000 edition, the first in a projected series, compiles our present information into three parts: Asia by Theme, Asia by Country and Region, and Indicators. For coming editions in this projected series we intend to include more themes and countries/regions, and continue to augment the associated data.

An increasingly vital task for solving environmental problems in Asia will unquestionably be the matter of how to provide for stronger cooperation and solidarity among Asia's environmental NGOs. It is our hope that the founding of this project will be the first step in achieving that.

(MIYAMOTO Ken'ichi, AWAJI Takehisa, TERANISHI Shun'ichi)

---

[2] See, for example: UNESCAP, *State of the Environment in Asia and the Pacific*, 1990 & 1995.

Cater Brandon and Ramesh Ramankutty, *Toward an Environmental Strategy for Asia*, World Bank Discussion Papers, No. 224, 1993.

David O'Conner, *Managing the Environment with Rapid Industrialization: Lesson from the East Asian Experience*, Development Centre of the OECD, 1994.

Foreword to the English-Language Version

This book is based on the Japanese book *Ajia Kankyo Hakusho 1997/98*, or "Asian Environment White Paper," which was published in December 1997 by Toyo Keizai Inc as the first in a series of NGO-perspective reports on the environment in Asia. This translation includes some new information and data, as well as some editorial changes.

Behind the project to compile and publish this "Asian Environment White Paper" series is the Japan Environmental Council (JEC), which was founded in June 1979 as a distinctive networking group comprising mostly researchers and experts. JEC's purposes include conducting interdisciplinary studies and research on environmental issues and policy, and making policy proposals. Although 1999 is JEC's 20th anniversary, its history can actually be traced as far back as 1963 to the formation of the Pollution Research Committee, an interdisciplinary research group that already has a history of over 35 years. JEC was organized around the nucleus composed of this committee's members and on the foundation prepared by their groundbreaking activities. Since its founding in 1979 JEC has planned and held annual conferences nearly every year at venues ranging throughout Japan, and in conjunction with these conferences it has sponsored national symposia open to the public. Additionally, for the nearly three decades since July 1971 JEC has edited and issued the quarterly magazine *Kankyo to Kogai (Research on Environmental Disruption*, published by Iwanami Shoten, Publishers), which is more or less JEC's official publication. Another JEC effort is the "J-JEC Environmental Book Series" published by Jikkyo Publishing Inc.

In November 1991 JEC became a membership organization, and by 1999 its register had grown to about 400 members who include university researchers and experts in a variety of disciplines, professionals, physicians, journalists, leaders of citizen movements around Japan, graduate students, ordinary citizens, and others.

With the arrival of the 1990s an intense awareness grew among JEC members concerning the importance of performing studies and research on the grave pollution and environmental problems becoming manifest especially in Asia, as well as on the kind of environmental policies needed to address them, and since that time JEC has labored to build a network leading to the formation of the Asian Environmental Council (AEC) while cooperating with independent researchers and experts working on the environmental issues and policy of Asian countries, and with a number of NGOs. One concrete initiative for this purpose is sponsoring the Asia-Pacific NGO Environmental Conference (APNEC), which has been held four times thus far. The first meeting was held in Bangkok in December 1991 with participants from eight countries; the second was in Seoul in March 1993; the third was held in Kyoto in November 1994; and the fourth was held in Singapore in November 1998 with participants from 15 countries. Using the foundation established by these steady network-building efforts, JEC in early 1995 launched a project to compile and publish this Asian Environment White Paper series with an NGO perspective. While the watchword "global environmental conservation starts in Asia" may be somewhat grandiose, it represents a strong feeling shared by everyone involved in the project, and it is the fundamental awareness with which we apply ourselves to this task.

Wide acclaim for *Ajia Kankyo Hakusho 1997/98*, the Japanese version of the first book in this series, has brought it to a fourth printing, and the compilation and editing of manuscripts for the second book — *Ajia Kankyo Hakusho 2000-2001* — are currently in progress.

As this English-language version goes to press, it is our sincere hope that, by winning readers in other parts of the world, this series will prove itself beneficial in advancing, even to a slight extent, the various initiatives directed at conserving the environment in Asia and across the globe as we enter the 21st century.

We would like to take this opportunity to express our gratitude to the many people who helped produce this English-language version.

First of all, our thanks go to Springer-Verlag Tokyo,

which graciously agreed to publish the book.

Our greatest thanks go to Mr. Rick Davis, who served not only as translator, but also doubled as editor for this version. It is no overstatement to say that this English-language version would not have been possible without his effort, cooperation, and volunteer spirit.

We are further indebted to the editors and writers named in the following list for their unstinting cooperation in checking the translated manuscripts (especially place names and personal names) time and time again despite their busy schedules. Also deserving mention here are Ms. Hirata Akiko, Mr. Oshima Kenichi, and Mr. Katayama Hirofumi who, as the unseen assistants behind this effort, generously dedicated themselves to organizing the manuscripts and performing the unbelievably large number of other clerical tasks that this English-language version entailed.

Finally, we would like to note that publication of this English-language version was made possible by generous grants from The Toyota Foundation, which provided strong support from the very outset of the project, as well as grants from the Japan Fund for Global Environment and the Consumers' Life Institute, Tokyo. We should like to express our profound gratitude to everyone at these institutions for their crucial support.

For the Japan Environmental Council *Ajia Kankyo Hakusho* Editorial Committee

August 1999

AWAJI Takehisa
TERANISHI Shun'ichi

Editors, Writers, Collaborators, and Assistants

Note: Names of East Asians are written according to East Asian custom with surnames first.

Editorial Advisors

HARADA Masazumi, Kumamoto Gakuen University

KIHARA Keikichi, Edogawa University

MIYAMOTO Ken'ichi, Ritsumeikan University

OKAMOTO Masami, Nihon University

SHIBATA Tokue, Center for Urban Studies, Tokyo Metropolitan University

UI Jun, University of Okinawa

Editorial Committee Members

ASUKA Jusen, The Center for Northeast Asian Studies, Tohoku University

AWAJI Takehisa, Rikkyo University

INOUE Makoto, The University of Tokyo

ISONO Yayoi, Tokyo Keizai University

ISOZAKI Hiroji, Iwate University

KOJIMA Nobuo, Attorney at Law

MATSUMOTO Yasuko, Science University of Tokyo, Suwa College

MIZUTANI Yoichi, Shizuoka University

NAGAI Susumu, Hosei University

TERANISHI Shun'ichi, Hitotsubashi University

UETA Kazuhiro, Kyoto University

YOSHIDA Fumikazu, Hokkaido University

Writers

AHYAUDIN bin Ali, University Sains Malaysia

AIKAWA Yasushi, The University of Tokyo, Graduate School

ASUKA Jusen, The Center for Northeast Asian Studies, Tohoku University

AWAJI Takehisa, Rikkyo University

CHANG Jung Ouk, Kyoto University Graduate School

CHEN Li-Chun, Yamaguchi University

CHEONG Deoksu, Osaka Kun-ei Women's College

HARADA Kazuhiro, JICA (Japan International Cooperation Agency) expert, Directorate General of Forest and Nature Conservation, Department of Forestry, Republic of Indonesia

HARADA Masazumi, Kumamoto Gakuen University

HISANO Shuji, Hokkaido University

HOSHINO Yoshiro, professor emeritus, Teikyo University

INOUE Makoto, The University of Tokyo

ISHIHARA Akiko, TRAFFIC Japan

ISONO Yayoi, Tokyo Keizai University

ISOZAKI Hiroji, Iwate University

JIN Song, China's Center for Sustainable Development, Peking University, Beijing

KAWAKAMI Tsuyoshi, The Institute for Science of Labour

KOJIMA Michikazu, Institute of Developing Economies, Japan External Trade Organization

KOJIMA Nobuo, Attorney at Law

KOJIMA Reietsu, Daito Bunka University

LEONG Yueh Kwong, Socio-Economic and Environmental Research Institute (Penang, Malaysia)

MARUTA Sayaka, Iwate University

MEENAKSHI Raman, Attorney at Law

MIYAMOTO Ken'ichi, Ritsumeikan University

MIZUTANI Yoichi, Shizuoka University

NAGAI Susumu, Hosei University

NAKAMURA Reiko, Ramsar Center Japan

OKUBO Noriko, Konan University

OSHIMA Ken'ichi, Takasaki City University of Economics

OTA Kazuhiro, Kobe University

SAKUMOTO Naoyuki, Institute of Developing Economies, Japan External Trade Organization

SEONG Won Cheol, Chukyo University

SHIBATA Tokue, Center for Urban Studies, Tokyo Metropolitan University

Sunee MALLIKAMARL, Chulalongkorn University

TACHIBANA Satoshi, The University of Tokyo

TERANISHI Shun'ichi, Hitotsubashi University

UETA Kazuhiro, Kyoto University

UEZONO Masatake, Shimane University

Uthaiwan KANCHANAKAMOL, Chiang Mai University

WANG Xi, Wuhan University

YOKEMOTO Masafumi, Tokyo Keizai University

YOSHIDA Fumikazu, Hokkaido University

YOSHIDA Hiroshi, Tokyo University of Agriculture and Technology

YOSHIDA Masato, Nature Conservation Society of Japan

Collaborators

Balan BALANI, The Consumers' Association of Penang

CHAFID Fandeli, Gadjah Mada University

CHUN Eung Hwi, National Council of YMCA of Korea

FUJINO Tadashi, Minamata Kyoritsu Hospital

FUJISAKI Shigeaki, Institute of Developing Economies, Japan External Trade Organization

HAN Gie-Yang, Ulsan Federation for Environmental Movement

ITO Shoji, The Tokyo Shimbun

JANG Won, Green Korea

KIM Bok Nyeo, Pusan Federation for Environmental Movement

KIM Jung Wk, Seoul National University

KOO Ja Sang, Pusan Federation for Environmental Movement

LEE Inhyun, Green Korea

LEE See Jae, Korean Federation for Environmental Movement

LI Jinchang, Jinchang Environmental Research, Beijing

MIZUNO Takeo, Attorney at Law

MORITA Tsuneyuki, National Institute for Environmental Studies, Japan

OTSUKA Kenji, Institute of Developing Economies, Japan External Trade Organization

PAN Ning, Interpreter

RHEE Jeong-Jeon, Seoul National University

SATO Yukihito, Institute of Developing Economies, Japan External Trade Organization

SHAM Sani, University Kebangsaan, Malaysia

SHAW Daigee, Academia Sinica, Taipei

Suraphol SUDARA, Chulalongkorn University

SUZUKI Katsunori, Environment Agency, Japan

TAKAHASHI Katsuhiko, AMR

TANI Yoichi, Solidarity Network Asia and Minamata

XU Houen, Department of Toxicology, Beijing Medical University

YAN Yang, Interpreter

YEH Jiunn-Rong, National Taiwan University

ZHANG Qide, Research Center for Environmental Sciences, Liaoning Province, China

Editorial Assistants

ANDO Makoto, Hitotsubashi University Graduate School

HIRATA Akiko, Hitotsubashi University

KATAYAMA Hirofumi, Hitotsubashi University

Contents

Part III   Indicators **121**

Part I   Asia by Theme

Chapter 1
Accelerated Industrialization and Explosive Urbanization

Part of the Onsan industrial complex in South Korea. Local residents have been subject to a semi-obligatory relocation program because of the grave pollution of the area.
Photo: Oshima Ken'ichi

1. East Asia at a Critical Juncture

During the two to three decades since the later half of the 1960s to the mid-1990s, Asia — especially East Asia — has achieved economic development of a rapidity unmatched anywhere else in the world. The World Bank focused on this in its 1993 policy research report (referred to below as "Bank report"), *The East Asian Miracle*,[1] in which it made the following statement.

"East Asia has had a remarkable record of high and sustained economic growth. From 1965 to 1990 the twenty-three economies of East Asia grew faster than all other regions of the world... Most of this achievement is attributable to seemingly miraculous growth in just eight economies: Japan; the "Four Tigers" — Hong Kong, the Republic of Korea, Singapore, and Taiwan, China; and the three newly industrializing economies (NIEs) of Southeast Asia, Indonesia, Malaysia, and Thailand."

Indeed, East Asia boasts a far higher average per capita GNP growth rate from 1965 to 1990 than other categories (Fig. 1; "East Asia" here includes all the low- and middle-income countries in East Asia and South Asia located in the eastern part of the region including China and Thailand, and in the Pacific). What's more, there is no doubt that Japan and the seven other economies in the World Bank quote above (termed "high-performing Asian economies," or HPAEs) have powerfully driven this economic growth. But there is perhaps dissent on whether to assess this process as the Bank report has done by praising it as a "marvelous record."

For example, MIT Professor Paul Krugman and others see this East Asian economic growth as little more than the result of increasing inputs of labor and capital (Krugman characterizes it as "input-driven growth"), and some entertain serious doubts that such growth can be sustained.[2] While these may be somewhat extreme views, now that we have passed the midpoint of the 1990s it is clear that East Asia's

growth as seen heretofore has arrived at a critical juncture, a fact symbolized by the chain-reaction economic crisis that began with the sudden drop in Thailand's currency in the summer of 1997.

Further, if we liberate our field of view from the framework of narrow assessments such as "economic performance" (for instance, how much economic growth has been attained under the measure of GNP) based on market economy indicators like those noted above, and widen it to include "environmental performance" that includes the reality of serious and steadily accumulating pollution and environmental damage, it would be hard to say that recent East Asian economic growth represents environmentally sustainable economic development. Especially if we think of what the 21st century will bring, it is unmistakable that, from an environmental perspective, it will be a time when people consider anew the very meaning of "quality of growth."

2. Review of Economic Growth in East Asia

Let us begin with a brief review of economic growth during the last few decades in countries (regions) representative of East Asia: the six HPAEs named in the Bank report, i.e., Japan, South Korea, Taiwan, Thailand, Malaysia, and Indonesia, plus China.

The economic growth rates of these seven economies from 1960 to 1996 (Table 1) show clearly that Japan, as the East Asian top runner, achieved rapid annual growth of about 10% (doubling the GDP in about seven years) from the 1960s to the early 1970s, but subsequently its growth rate dropped to half or even lower, partly as a result of the two oil shocks. But then South Korea and Taiwan achieved high annual growth of 8 to 10% from the late 1960s to the late 1980s, after which Thailand, Malaysia, Indonesia, and China charged up from behind and attained rapid growth of 8 to 10% from the late 1980s to the mid-1990s. So in effect, from the 1960s to the mid-1990s a wave of rapid growth swept over East Asia, beginning with Japan, then to the representative NIEs of South Korea and Taiwan, then to the representative ASEAN countries of Thailand, Malaysia, and Indonesia, and finally to China, especially the South China coastal region.[3] Fast growth resulted in skyrocketing per capita GDP levels in these economies during a mere two to three decades (Table 2). For that reason the Bank report, and indeed the whole world, has focused especially on East Asia as the dazzling symbol of rapid Asian progress.

However, that economic growth was also a process that

Fig. 1 Average Growth Rate in Per Capita GNP, 1965-1990 (by world regions)

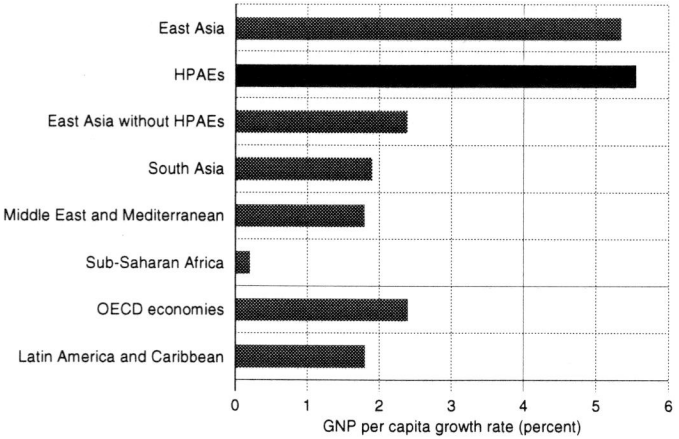

GNP per capita growth rate (percent)

Source: World Bank.

[1]   World Bank, *THE EAST ASIAN MIRACLE: Economic and Public Policy — A World Bank Policy Research Report*, Oxford University, 1993.

[2]   Krugman's view was much discussed after appearing in the November and December 1994 issues of *Foreign Affairs*, but there was a great deal of criticism and many refutations from experts on Asian economies. One was Komiya Ryutaro and Yamada Yutaka, ed., *Higashi Ajia no Keizai Hatten — Seicho wa Doko made Jizoku Suru ka* (Economic Development in East Asia — How Long Will Growth Continue?), Toyo Keizai Inc, 1996.
[3]   The general view is that this wave will henceforth sweep over China's interior and India, then over Vietnam and other countries of Indochina.

swallowed up these economies in the kind of severe pollution and environmental damage that occurred in Japan during its own rapid growth in the 1960s.

Following are a few of the basic characteristics of this process.

3. Accelerated Industrialization

The first characteristic of this economic growth in East Asia is that it involved a dramatic transformation of the industrial structure that could be called "accelerated industrialization" (i.e., industrialization that took several generations

Table 1 Growth Rates of Seven East Asian Economies

(Calendar year, %)

	Japan	South Korea	Taiwan	Thailand	Malaysia	Indonesia	China
1960	13.3	1.2	6.3	9.2	-	0.7	-
61	11.9	5.8	6.9	5.2	-	4.2	-
62	8.6	2.1	7.9	7.8	-	-0.6	-
63	8.8	9.1	9.4	8.0	-	-0.4	-
64	11.2	9.7	12.2	6.7	-	4.8	-
65	5.7	5.7	11.1	7.8	-	0.0	-
66	10.2	12.2	8.9	11.5	-	2.3	-
67	11.2	5.9	10.7	8.1	-	2.3	-
68	11.9	11.3	9.2	8.1	-	11.1	-
69	12.0	13.8	9.0	7.9	-	6.0	-
70	10.3	8.8	11.4	10.6	-	7.5	-
71	4.4	9.2	12.9	4.9	7.1	7.0	-
72	8.4	5.9	13.3	4.1	9.4	9.4	-
73	8.0	14.4	12.8	10.0	11.7	11.3	-
74	-1.2	7.9	1.2	4.2	8.3	7.6	-
75	3.1	7.1	4.9	4.9	0.8	5.0	-
76	4.0	12.9	13.9	9.3	11.6	6.9	-
77	4.4	10.1	10.2	9.9	7.8	8.8	-
78	5.3	9.7	13.6	10.5	6.7	7.8	11.7
79	5.5	7.6	8.2	5.3	9.3	6.3	7.6
80	2.8	-2.2	7.3	4.8	7.4	9.9	7.9
81	3.2	6.7	6.2	6.0	6.9	7.9	4.4
82	3.1	7.3	3.6	5.3	5.9	2.3	8.3
83	2.3	11.8	8.5	5.5	6.3	4.2	10.4
84	3.9	9.4	10.6	5.8	7.8	7.0	17.3
85	4.4	6.9	5.0	4.6	-1.0	2.5	13.5
86	2.9	11.6	11.6	4.9	1.0	5.9	8.8
87	4.1	11.5	12.7	9.5	5.4	4.9	11.6
88	6.2	11.3	7.8	13.3	8.9	5.8	11.3
89	4.8	6.4	8.2	12.2	9.2	7.5	4.1
90	5.1	9.5	5.4	11.6	9.7	7.2	3.8
91	3.8	9.1	7.6	8.4	8.7	7.0	9.2
92	1.0	5.1	6.8	7.9	7.8	6.5	14.2
93	0.3	5.8	6.3	8.3	8.3	6.5	13.5
94	0.6	8.6	6.5	8.8	9.2	7.5	12.6
95	1.4	8.9	6.0	8.6	9.5	8.2	10.5
96	3.6	7.1	5.7	6.7	8.2	7.8	9.7
1960s	10.4	7.6	9.1	8.0	-	3.0	-
1970s	5.2	9.3	10.2	7.3	8.0	7.7	9.6
1980s	3.8	8.0	8.1	7.1	5.7	5.7	9.9
1990s	2.2	7.7	6.3	8.6	8.8	8.8	11.6

Notes: 1. GDP growth rates (real, based on each country's own currency).
            2. Dashes mean no data available.
Source: Prepared from the May 1997 *Asian Economy 1997*, edited by the Economic Planning Agency's Research Bureau.

6

Table 2 Per Capita GDP in Seven East Asian Economies

(Calendar year, US$)

	Japan	South Korea	Taiwan	Thailand	Malaysia	Indonesia	China
1960	477	115	-	97	275	-	-
65	932	106	223	131	312	-	-
70	1,967	272	386	194	382	77	-
75	4,475	599	962	355	784	225	-
80	9,146	1,643	2,325	693	1,785	491	302
85	11,282	2,311	3,223	755	1,994	531	291
90	24,276	5,917	7,870	1,527	2,415	590	342
95	41,045	10,037	12,213	2,750	4,337	1,039	584

Notes: 1. Dollar conversions performed using average annual rates from the IMF's *International Financial Statistics*.
2. Dashes mean no data available.
Source: Same as Table 1.

in the developed nations was achieved in just one generation).

For example, the 1948 tragedy of dividing the Korean peninsula cut South Korea completely off from the Korean industrial sector that had until that time been concentrated mainly in the north, which meant that South Korea started as a very poor agricultural country in which nearly 90% of the people were small farmers. But thanks to subsequent farmland reform and national economic development, the industrial sector accounted for a rapidly increasing proportion of the GDP, rising to 25% in 1965, and to 45% in 1990 owing to emphasis on the heavy and chemical industries that began in the 1970s. On the other hand, the agricultural sector dropped precipitously from 39% in 1965 to a mere 9% 25 years later in 1990. Taiwan likewise began as an agricultural society; during these years its industrial sector grew rapidly from 29% in 1965 to 43% in 1990, while its agricultural sector plummeted from 27% in 1965 to 5% in 1990. Even in countries like Thailand and Indonesia, which are still basically agricultural societies, their respective industrial sectors have risen quickly from 23% and 13% in 1965 to 39% and 40% in 1990, while their agricultural sectors declined from 43% and 51% in 1965 to 12% and 22% in 1990, likewise dropping very quickly.[4]

Needless to say, even the Western industrialized countries, which set out on the road to industrialization very early, went through the same process in which importance gradu-

ally shifted from agriculture to industry as their economies grew, but it happened much more slowly than in the East Asian economies discussed here. Postwar Japan went through this process at about twice the speed of Western industrialized countries during its rapid economic growth phase, but within the last two or three decades the other East Asian economies discussed here have achieved industrialization at a very high rate, outpacing even that of Japan. This we might called "accelerated" development.

4. Explosive Urban Growth

The second characteristic is that this process of "accelerated industrialization" occurred in conjunction with explosive urban growth that was magnified by the impoverishment of rural areas suffering under the onus of huge populations. For example, during the 20 years from 1950 to 1970 in Japan the urbanization ratio jumped from 50.3% to 71.2%, then slowed somewhat (Table 3). But urbanization in South Korea has proceeded even faster than in Japan, increasing from 21.4% in 1950 to 40.7% in 1970, and then after another two decades to 73.8% in 1990. Even in Thailand and Indonesia, which on the whole are still at relatively low urbanization levels, the excessive concentration of population in urban centers like Bangkok and Jakarta, and the rate of urbanization there, are truly explosive.

Meanwhile, world population has risen sharply since 1950, when it was about 2.5 billion, to about 3.7 billion in

---

[4] The foregoing figures are from the World Bank's *World Development Report 1992*.

Table 3 Urbanization Ratios in Seven East Asian Economies

(%)

	Japan	South Korea	Taiwan	Thailand	Malaysia	Indonesia	China
1950	50.3	21.4	-	10.5	20.4	12.4	11.0
1970	71.2	40.7	-	13.3	33.5	17.1	17.5
1990	77.2	73.8	(75.9)	18.7	49.8	30.6	26.2
1994	77.5	80.0	(76.6)	19.7	52.9	34.4	29.4
2010 (predicted)	80.6	91.4	-	27.4	64.4	49.7	43.0

Notes: Owing to differing conceptions of cities depending on country, urbanization ratio (i.e., urban population/total population) here is based on urban population estimates (prepared by the UN Population Bureau with the cooperation of the UN Statistics Bureau) in urban agglomerations instead of urban areas proper. Thus there are no figures for Taiwan, whose parenthetical values are taken from *Statistical Yearbook of the Republic of China*.
Source: United Nations, *World Urbanization Prospects: The 1994 Revision*, 1995.

1970 and about 5.7 billion in 1995. Especially Asia (consisting here of "East Asia," "Southeast Asia," "South Central Asia," and "West Asia" in UN population statistical categories) soared from about 1.5 billion in 1950 to 3.46 billion in 1995, already accounting for 65% of world population. This huge population is one major characteristic of Asia, and is one of the factors underlying this explosive urbanization in Asian countries and regions. Further, accelerated industrialization is a major cause triggering Asia's rapid urbanization.

5. Asia Joins the Mass Consumption Society

The third characteristic of Asia's economic growth is that it is coming up from behind the industrialized countries and, with an even greater speed, bringing a mass consumption lifestyle to Asian peoples. On one hand, the spread of this lifestyle is a manifestation of the growing middle class, which is itself a product of higher incomes, but another cause is the early appearance of the consumer loan in Asian countries. It is a fact that the increase in goods like automobiles and electric appliances outstrips the rise in income. Professor Kojima Reietsu (Daito Bunka University) notes that Japan joined the mass consumption society at the end of the 1950s and was close behind the U.S. a decade later. Other economies made their entries at subsequent times: Hong Kong at the beginning of the 60s, Taiwan at the end of the 60s, South Korea in the early 70s, and China's major coastal cities at the beginning of the 80s. This means that Asians have become full-fledged members of the energy-intensive, mass-consumption, mass-refuse society.

6. Mounting Costs of Pollution and Environmental Problems

The three characteristics given above — (1) accelerated industrialization, (2) explosive urbanization, and (3) joining the mass-consumption society — are the primary factors behind the unprecedented rapid growth of East Asia's economies. From another perspective, they have caused the structural exacerbation of East Asia's compound pollution and environmental damage, and will make them more serious in the future. As the following chapters will demonstrate, East Asia as a whole shares the following difficult challenges, which are manifestations of composite problems such as (1) combined industrial and urban pollution, (2) combined traditional and modern problems, and (3) combined domestic and international factors. Asian countries and regions share the task of how to address these difficult challenges.

7. Summation: Qualitative Shift in Economic Growth

East Asia now finds a pressing need for a fundamental shift in the very quality of its economic growth. Asia has arrived at a time when it must break new ground in an unprecedented search: the challenge of building a new eco-compatible socioeconomic system. This need arises not only because of down-to-earth realities in today's Asia, but also because of the urgent and inescapable task of conserving the global environment in the coming century.

(TERANISHI Shun'ichi, OSHIMA Ken'ichi)

Essay   Expanding Use of Nuclear Power and Increasing Risks in Asia

Since the 1986 Chernobyl nuclear accident there has been a distinct worldwide ebb in nuclear power, but East Asian countries and regions have instead been promoting it aggressively for the avowed purposes of securing energy supplies and addressing global warming. As of September 1997 Japan had 53 operating power plants, South Korea 12, Taiwan six, and China three. That August there was a groundbreaking ceremony in North Korea for the construction of two nuclear plants with KEDO (Korean Peninsula Energy Development Organization) assistance. Indonesia, Vietnam, Thailand, and other Southeast Asian countries and regions are planning or considering the contruction of nuclear power stations. If all plans come to fruition there will be at least 120 nuclear plants in 2010, giving East Asia the highest density of nuclear power plants in the world. Behind this spread of nuclear power in Asia are the aggressive nuclear plant export policies of governments and plant makers in the developed nuclear countries. Because plant orders are slumping in these countries, they seek to preserve their own plant makers' viability by exporting as they help look for sources of funding. For example, China's Daya Bay nuclear plant is funded by an international syndicate loan that includes the World Bank, while equipment and technology come from France and Britain. Funding for the Qinshan Nos. 4 and 5 reactors, whose turbines and generators are supplied by Hitachi, Ltd., came from the U.S. Export-Import Bank ($1.8 billion for 67% of the foreign investment), and the Japan Export-Import Bank and Tokyo Mitsubishi Bank ($280 million for 10.5%).

Meanwhile, there is an increasing accident risk owing to the aging of reactors, mainly in Japan, and fever-pitch reactor construction in China. Despite continuing nuclear plant accidents in Japan, periodic inspection times are shortened to improve reactor efficiency, and there are also plans to use the more dangerous MOX fuel. In the case of China, there are concerns about not only the safety of the intermediate components in China's small self-developed nuclear plant (Qinshan No. 1, 300 MW), but also the risk entailed in operating and managing plants supplied by different countries, including China, Canada, France, and Russia. Indonesia intends to build a nuclear plant on Java, which is subject to volcanos and frequent earthquakes. In conjunction with this growing number of nuclear plants, there are increasing risks associated with the transport and storage of nuclear fuel and wastes. Additionally, Russia's illegal dumping of radioactive wastes in the Sea of Japan, which came to light in 1993, Taiwan's contracting North Korea to store its radioactive wastes, and other events augment the possibility of radioactive waste accidents in the Sea of Japan.

(CHANG Jung Ouk)

Chapter 2
The Accelerating Car Culture

Bicycles, the citizens' transportation, and rapidly increasing motor vehicles in a Shanghai intersection. It will not be long when, like Bangkok, bicycles give way to motorbikes, and then to automobiles.
Photo : Isono Yayoi

1. Motor Vehicle Pollution in Bangkok

1-1 High-Concentration Airborne Particulates

As Asia undergoes swift economic growth and urbanization, Bangkok offers a prime example of a typical major city that is worsening its pollution, especially that from motor vehicles. The increase in Bangkok's motor vehicle fleet is particularly rapid. For instance, motorcycles and other registered vehicles multiplied from 600,000 in 1980 to 2.7 million in 1993, a 4.5-fold increase in a mere 10-odd years. Because that number accounts for one-fourth of all Thailand's motor vehicles, Bangkok's streets are jammed not just in the morning and evening, but all day long. The fleet has since continued to increase quickly, with 600,000 new vehicles sold in 1996. A difference with other developed countries is that 46% of Bangkok's vehicles are two-wheeled, and at least 90% of those have two-stroke engines, which means that the city's traffic jams and vehicle exhaust fumes generate appalling air pollution. Serious pollutants include airborne particulates, hydrocarbons (HCs), and lead. A recent survey showed that Bangkok's 24-hour average in 1994 for particulates exceeded Thailand's 330 $\mu g/m^3$ standard at all roadside monitoring stations. While the average value is 350 $\mu g/m^3$, individual readings ranged from 50 to 1,130 $\mu g/m^3$. Further, the standard was exceeded 138 out of 261 monitoring days, which is 53% of the time. By weight, 40% of particulates are from diesel engines, 40% are road dust, and 20% are from factories and other business facilities. Especially the 24-hour maximum of 1,130 $\mu g/m^3$ in 1994 (the previous year's maximum had been 1,500 $\mu g/m^3$) was a shock to the Thai government, which hurriedly convened the Transport Ministry, police, chamber of commerce, and other organizations to consider comprehensive remedial measures. In the summer of 1995 the seriousness of the situation forced them into an agreement to lower the maximum to 800 $\mu g/m^3$ by 1997, 500 by 2000, and below 300 by 2001, and to develop short- and long-term action plans.

1-2 Traffic Congestion and the Social Cost of the Automobile

Already the World Bank and others have tried measuring as social cost the chaos and pollution of motor vehicle traffic in Bangkok. They have shown that if air pollution by airborne particulates and leaded gasoline were reduced by 20%, annual benefit owing to decreased health damage, i.e., recovery from diseases and a lower death rate, would be the equivalent of US$1 to 1.6 billion, and $300 million to 1.5 billion, respectively. In 1995 the Research Department of the influential Thailand Farmers Bank released the following calculations on motor vehicle social cost, which were of great interest in Thailand and abroad. Each day 1.5 million motor vehicles enter central Bangkok, making for an average speed 15-20 km/h, but if the average traffic speed when

flowing smoothly is assumed to be 25 km/h, the resulting loss in fuel cost is 12.36 billion baht. Additionally, the average commuter living within 30 km of central Bangkok wastes 86 minutes each day in commuting time because of traffic jams. If 50% (or 20%) of this time were used in productive activity, the value obtained would be 37.26 billion baht (or 14.9 billion baht).

Thai Ministry of Health statistics say there are about 1.9 million people with respiratory difficulties in Bangkok, accounting for one in four Bangkok citizens. The Farmers Research Center emphasizes that this is a reliable figure based on Health Ministry and hospital studies. Interviews with physicians show that 80 to 90% of these people are victimized by air pollution, and most of the air pollutants are from motor vehicles. If these 1.9 million people visit the hospital 10 times a year and pay 500 baht on each visit, that comes to about 9.5 billion baht in medical care costs. Total motor vehicle social costs of petroleum consumption, time benefit, and health damage amount to 59.13 billion baht annually, or 28.44 baht per person per day. Even a lower estimate comes to 36.77 billion baht, whose estimated effect on Thailand's overall economy is to lower the growth rate 2.5%, investment 1.86%, and consumption 1.65%. These results led the Farmers Research Center to propose the construction of main roads and mass transit systems, as well as suppressing growth of the automobile fleet, limiting motor vehicle entry into the central city as in Singapore's area licensing scheme, instituting a flex-time scheme, and other policies to control traffic demand.

Bangkok is now internationally well known for its serious traffic jams and motor vehicle exhaust, but social cost calculations by a major Thai bank have shown that the worsening motor vehicle pollution in recent years is adversely affecting the productivity of Thailand's economy.

Under Bangkok's inadequate urban planning the Thai government and the Bangkok Metropolitan Administration (BMA) plan road construction independently of one another. Owing to conversion of canals to roads and the large landowner system, the city is full of one-way or dead-end alleyways called *soi*. Cars therefore descend on the few main streets, and traffic increases because drivers must take roundabout routes to their destinations. Especially the lack of urban planning has created an imbalance between business and residential districts, making it very difficult to provide public transportation. In the construction of roads and public transportation Thailand tends to depend on foreign aid or private funds for build-operate-transfer (BOT) schemes, so plans do not always proceed smoothly. Further aggravating the exhaust problem are the three-wheeled taxis called *tuk-tuk*, and the poorly maintained two-wheel and used vehicles that crowd the streets, thus creating a grave situation of harm by air pollution, including damage to the health of the many police officers directing traffic.

Table 1 Motor Vehicle Market Trends in Asia

(Thousands of vehicles)

	Domestic demand				Production			
	83	88	93	88-93 growth rate	83	88	93	88-93 growth rate
Thailand	118	142	456	26.3	109	157	419	21.7
Indonesia	152	153	214	6.9	125	159	203	5.0
Malaysia	122	69	154	17.4	118	98	180	12.9
Philippines	49	17	84	37.6	42	17	*84	37.6
Singapore	39	31	43	6.8	-	-	-	-
ASEAN total	480	412	951	18.2	394	431	886	15.5
China	**252	645	1,342	15.8	239	593	1,280	16.6
Taiwan	153	398	568	7.4	150	281	404	7.5
S. Korea	195	523	1,437	22.4	221	1,083	2,050	13.6
India	158	314	381	3.9	154	312	371	3.5
Nine-country total	1,238	2,292	4,679	15.3	1,158	2,700	4,991	13.1

Notes: * The Philippines' production in 1993 was assumed to be equal to domestic demand.
       ** China's 1983 domestic demand was production plus imports.
Source: Okada Mitsuhiro, "Rapid Growth of Asia's Automobile Industry," *JAMAGAZINE*, Vol. 29, November 1996.

2. Motor Vehicles and Urban Air Pollution

2-1 Steadily Increasing Motor Vehicle Use

The seriousness of motor vehicle pollution in Bangkok is common to varying extents in Asia's major cities because rapid economic growth spurs the growth of big cities, which in turn brings more vehicles. Indeed, annual growth in domestic automobile sales during the decade from 1983 to 1993 in the five ASEAN countries of Thailand, Indonesia, Malaysia, the Philippines, and Singapore was in the double digits, at 18.2%, a contrast with the industrialized countries where growth has all but stopped or is very low (Table 1). The growth in sales for nine economies with China, South Korea, Taiwan, and India added is 15.3%. These increasing automobile sales in Asian countries mean that production is also rising. Those same nine countries have a production growth rate of 13.1%, and a number of studies suggest that their motor vehicle demand will continue to rise from 4.68 million vehicles in 1993 to about 8 million in 2000 to over 10 million in 2005. According to estimates, in 2005 vehicle demand in those nine countries including Japan will be about 18 million, far surpassing the North American and Western European markets.

However, while this increase in vehicle demand signifies greater ownership engendered by higher incomes in Asia, this phenomenon generally means a swift increase in ownership from several dozen vehicles to 150 or 200 per 1,000 population. Past trends in developed countries' automobile societies show a general tendency in which, once ownership has attained about 10 vehicles per 1,000 population, it gradually picks up speed, reaches 50 vehicles in 10 years, and accelerates even more and rises to about 200 vehicles in seven or eight years, after which the rate gradually slows. From this we can see that as China is now in the initial stage of about 10 vehicles per 1,000 people, the increase in ownership will likely quicken from now on. South Korea, on the

other hand, already has about 180 vehicles per 1,000 people, so its rate of increase may presently begin to slow. Applying this empirical rule on vehicle ownership to Asian countries leads one to inescapably anticipate that, expect for Japan and South Korea, motor vehicles will take over Asian cities with ever greater speed. Still, as the typical example of Bangkok shows, even at the present stage the automobile society is causing serious harm, and effective remedial measures will not come easily.

2-2 Worsening Air Pollution in Major Asian Cities

In conjunction with rising car ownership in Asia, exhaust-caused air pollution has now become quite serious in Asia's major cities. A WHO/UNEP 1992 report, *Urban Air Pollution in Megacities of the World*, estimates that in 2000 the world will have 20 cities with populations of 10 million or more, and 11 of them (Bangkok, Beijing, Bombay, Calcutta, Delhi, Jakarta, Karachi, Manila, Seoul, Shanghai, and Tokyo) are in Asia. The report anticipates that rapid evolution into big cities will occur especially in Asian developing countries. This rapid urbanization will accelerate rising vehicle ownership, but under present circumstances air pollution in Asia's big cities is already very serious, especially that by particulates, lead, SO_2, ozone, and other pollutants. Twelve of the world's 20 megacities have particulate pollution over twice that of the WHO guideline, and 10 of them are major Asian cities (Fig. 1). And while Karachi has lead pollution over twice that of the WHO guideline, three cities with concentrations under twice the guideline but nevertheless high are Bangkok, Jakarta, and Manila (Fig. 2). Two cities over twice WHO's SO_2 guideline are Beijing and Seoul (Fig. 3). With respect to ozone, Tokyo is among the cities over two times the WHO guideline (Fig. 4).

As this shows, the already serious state of air pollution in Asia's major cities makes it impossible to ignore real health damage.

12

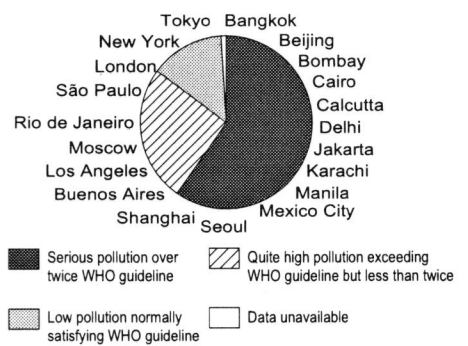

Fig. 1 Airborne Particulate Levels in 20 Major World Cities

Legend:
- Serious pollution over twice WHO guideline
- Quite high pollution exceeding WHO guideline but less than twice
- Low pollution normally satisfying WHO guideline
- Data unavailable

Source: WHO/UNEP, *Urban Air Pollution in Megacities of the World*, 1992, Blackwell.

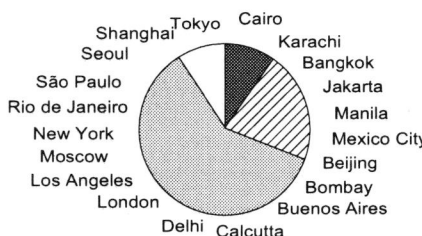

Fig. 2 Lead Levels in 20 Major World Cities

Legend: Same as Fig. 1.
Source: Same as Fig. 1.

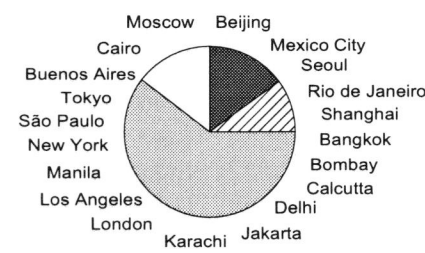

Fig. 3 SO₂ Levels in 20 Major World Cities

Legend: Same as Fig. 1.
Source: Same as Fig. 1.

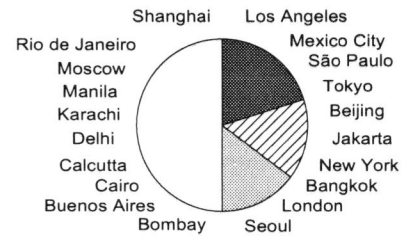

Fig. 4 Ozone Levels in 20 Major World Cities

Legend: Same as Fig. 1.
Source: Same as Fig. 1.

With lead, for example, WHO's standard is an annual average of 0.5-1 $\mu g/m^3$, but in Jakarta the concentration is 1.26 $\mu g/m^3$ along streets (1993-1994; in particular it was 1.65 $\mu g/m^3$ at Pasar Baru), making the risk of lead-induced health damage high. A study by Professor Umar Fahmi Achmadi at the University of Indonesia[1] showed that lead concentration in the blood of Jakarta slum residents was 0.928 $\mu g/l$, 14 times higher than that of people living in areas with clean air. The study points out health damage including lower labor productivity and lower intelligence of children (WHO observes the possibility that concentrations over 0.2 $\mu g/l$ may damage health). Note also that Jakarta gasoline is leaded in the world's highest category, at 0.6 to 0.73 g/l (1987 level), making it necessary for Jakarta to quickly introduce unleaded gasoline.

Thailand is a recent success story in phasing out leaded gasoline. In 1984 Thailand's gasoline contained 0.84 g/l of lead, which declined to 0.15 g/l in 1992. In 1994 regular gasoline was totally lead-free, and in 1996 so was premium. Lead is added to gasoline to raise its octane rating, but eliminating lead causes the problem of increasing hydrocarbon emissions. At the same time, gasoline-powered vehicles burning leaded gasoline cannot use three-way catalytic converters or other emission control devices, whose use was allowed by eliminating lead.

3. Challenges of Motorization: The Perspective from Japan's Postwar Experience

How widespread the ownership of motor vehicles becomes in any one country depends on factors including the increase in vehicle purchasing power due to rising GDP and incomes, and the rate and attributes of urbanization, but it also depends on the infrastructural readiness of cities to accommodate vehicular traffic. In postwar Japan, vehicle ownership began rising quickly during the latter half of the 1950s, and legislation to accommodate motor vehicles was part of the process (Table 2). While this is so of all countries, Asian countries will now have to provide infrastructure as their driving populations rise. This will include road improvements, building of parking lots, expressway construction, road signs, and developing and improving traffic safety measures, as well as creating vehicle inspection and insurance systems. Below we shall examine these requirements in the light of Japan's postwar experience.

3-1 Traffic Congestion and Road Construction

It was around 1956 when Japan switched from coal to oil and began full-blown road construction to accommodate the rising car culture. In that year the World Bank's Ralph

---

[1] "Epidemiological Perspective of the Impact of Air Pollutants in Cities of Indonesia," *Proceedings of Issues of Using Leaded and Unleaded Gasoline*, August 2, 1995.

W. Watkins Road Survey Commission visited Japan and delivered a report well-known for its statement that Japan had no roads, only right of ways slated for construction. As a matter of fact, at that time only 35% of national highways had been improved (meaning at least 5.5 meters wide, allowing two lanes of traffic), for a total length of only 8,450 km. Japan steadily pursued construction and improvement until in 1994 87% had been improved, for a total length of over 46,000 km. And while high-speed and heavy vehicles find driving on gravel and muddy roads difficult, in 1955 only 17.2% of national highways were paved. In 1994 this was 98.4% (86.4% even when excluding light-traffic pavement), so nearly all main roads are now paved.

Generally when road construction cannot keep pace with soaring vehicle ownership, the quickly ensuing consequences are traffic jams, increasing traffic accidents, and other turmoil, especially in large urban areas. Even in the vicinity of Indonesia's capital city of Jakarta, which has a total of 5,683 km of roads including 1,334 km of national roads, the road length per automobile is already a mere 5.7 meters, which is equivalent to Tokyo's 5.5 meters (as of 1992). In China, except for places such as Beijing's Tiananmen Square where automobile and bicycle traffic are fairly well separated, chaos reigns on ordinary streets where automobiles make their way indiscriminately through crowds of bicycles. On the other hand there are places like Seoul where the waves of motor vehicles make bicycle riding dangerous. Calcutta's streets have a jumbled crush of human-drawn carts, pedestrians, and animals. Especially in places like Howler Bridge over the Huguri River in Calcutta, where all this traffic concentrates, traffic is always congested because in this area it can pass only here. India still has relatively few motor vehicles, and passenger cars are used only by a small privileged class of high-income people; when they pass occasionally they sound their horns clamorously and everyone on the street gets out of the way to allow passage. While this somehow gets India by now, it is possible that traffic will grind to a halt as the streets gradually carry more automobiles. In Bangkok about 20 years ago, when vehicle ownership began to rise, the whole city was a blaring cacophony because drivers all honked their horns. Thailand is a nation of devout Buddhists, and every day in Bangkok citizens would gather at the great famous temples for meditation and Buddhist services amid quietude, but the sound of traffic totally disturbed this.

In postwar Japan specific funding for road and bridge construction and maintenance came from automobile-related tax revenues including the gasoline tax and diesel delivery tax, which were created under the 1958 Road Improvement Special Account Law. Passed around the same time were the Japan Highway Public Corporation Law and the Metropolitan Expressway Public Corporation Law, under which roads were built using funds from the low-interest, long-term fiscal investment and loan program. The government also created a system for tolls to recover funds spent on road construction, which proceeded quickly owing to these measures. Thanks to these funding mechanisms road investment swelled rapidly, until in the recent Eleventh Five-Year Road Construction Plan (1993-1997) it attained a colossal total of 76 trillion yen. Japan's way of funding road construction with a specific fund from automobile-related taxes has indeed served a purpose by promoting the construction of roads, but presently road building no longer makes for greater efficiency and has brought about fiscal inflexibility. Another problem of this system is that it cripples the overall transportation policy balance by encouraging only motor vehicle traffic. Unlike Japan, Asian developing countries have adopted foreign aid and BOT schemes for road construction, but these too involve problems that will be discussed below.

3-2 Inspection Systems and Poorly Maintained Vehicles

Even with the widening of existing roads and new road construction, allowing more traffic to pass smoothly and

Table 2 Development of Motor Vehicle-Related Policy in Japan

Year	Laws	Vehicle production (thousands)	Road investment (billions of yen)	People injured in accidents (thousands)
1955	Local Road Tax Law		62.3	
1956	Ralph W. Watkins Road Survey Commission visits Japan Japan Highway (Public) Corporation established	111	74.5	102
1957	Law for Construction of Arterial Motorways for National Land Development,Parking Place Law	182	110.8	125
1958	Emergency Financial Law for Road Improvement Road Improvement Special Account	188	138.1	185
1959	Metropolitan (Tokyo) Expressway Corporation established	262	175.9	231
1960	Road Transport Law	481	211.3	289
1965	Nagoya-Kobe Expressway opened	1,876	699.1	426
1966	Emergency Law for Traffic Safety Facilities	2,286	868.6	518
1970	Traffic Safety Measures Basic Law Environment-related legislation	5,289	1,597.9	981

Sources: Ministry of Construction, *Road Pocketbook* 1997, p. 6.
  Nikkan Automobile News, *Automobile Industry Handbook*, 1996 edition.

safely involves still other requirements. First of all, a country has to have a driver's license system, driving schools, and other institutions to ensure that drivers are capable. As of the end of 1994 Japan already had as many as 67.2 million licensed drivers (53.7 people in 100). Other Asian countries will also need to set up driver's license testing systems and the like to accommodate their growing vehicle fleets. Also important will be passing traffic laws, instituting penalties for offenders, and other infrastructural arrangements. Second, they will need to keep dangerous defective vehicles off their roads with vehicle registration and inspection systems. One concern is that developing countries especially have many dilapidated used vehicles for which there are no institutional provisions for repairs and inspections. Catalytic converters and other exhaust-control measures are inadequate. Bangkok's used automobiles and motorcycles number at least 1 million each and have high fuel-air ratios owing to poor maintenance. Each vehicle thus has unusually high carbon monoxide (CO) emissions, as well as high HC emissions owing to poor combustion.

Asian countries are now working on a variety of measures to strengthen controls on used vehicles or to establish and enhance vehicle inspection systems. In Bangkok, for example, many taxis and *tuk-tuk* have begun switching to LPG fuel, and all buses and trucks must submit to an exhaust emission test as part of their annual vehicle inspections. Furthermore, recently passenger cars and motorcycles in use more than 10 years and seven years, respectively, must undergo exhaust tests during their annual vehicle inspections. Private vehicle inspection garages are licensed to perform these inspections. At the same time, it is not clear what penalty provisions or binding improvement requirements have been instituted, so one cannot be sure how effective these inspections actually are.

Taiwan's initiatives in particular are ahead in this respect. Motorcycles qualifying as low-pollution vehicles test at under 7,000 ppm for HCs and under 3.5% for CO. Those passing tests emit 7,000-9,000 ppm HCs and 3.5-4.5% CO, while motorcycles with over 9,000 ppm HCs and over 4.5% CO do not pass. An April 1995 trial inspection covering 130,686 vehicles failed 27%. While Taiwan has yet to establish penalties and require maintenance for disqualified motorcycles, the authorities are developing an effective inspection system using magnetic cards and other means. Additionally, as of the end of 1993 Taiwan had a total of 15,190,000 motor vehicles. Of that total, 10,950,000 were motorcycles, a figure which shows that they are the principal mode of transportation. The proportions of Taipei's air pollution arising from motor vehicles are estimated at 95% for CO, 99% for NOx, 66% for HCs, and 12% for PM_{10} (the remainder being from factories and other fixed sources). Taiwan now also performs inspections for particulate emissions of diesel-powered vehicles (buses and trucks).

3-3 Traffic Safety Infrastructure

Even if a country has good drivers and well-maintained vehicles, a rapid increase in the vehicle fleet will translate into rapid increases in traffic accidents and victims. As Table 2 shows, the swift climbs in automobile production and road investment in Japan during the latter half of the 1960s brought about a sharp rise in accident victims, so that in 1970 the number of accident-caused deaths attained 16,765 (this figure is close to 20,000 if instead of deaths within 24 hours we use the international definition of death within one month). Since the 1966 Emergency Law for Traffic Safety Infrastructure, therefore, enhancing such infrastructure has been an urgent priority in Japan, where, as of the end of 1994, there were 2,232,166 road lighting fixtures, 129,948 km of guard rails, 1,874,050 road signs, and 10,483 locations with pedestrian overpasses (this number actually shows Japan's backwardness because it means that Japan gives vehicles priority over people). Almost all of these signs and other safety infrastructure have appeared since the 1970s, and other Asian countries will likewise have to quickly provide such infrastructure to accommodate the swift growth of their vehicle fleets. They will also need vehicle insurance to cope with traffic accidents.

3-4 Financing Urban Infrastructure

From now on it will be important in Asia to constrain the use of automobiles as much as possible while developing rail-based mass transit (streetcars, subways, monorails, new transit systems, etc.) to substitute for automobiles to the greatest possible extent in accordance with the peculiarities of each city. In that respect there is much to learn from Singapore, which restricts motor vehicle use in all forms and gives priority to public transit. Beijing and Calcutta are already using subways. Of course at present the former has two lines and the latter only one, so the distance covered is too short to cope with their burgeoning urban populations, and in particular they do not help solve the problem of traffic from city outskirts. Seoul has four subway lines, but they are not adequate as a substitute for its rapidly swelling motor vehicle fleet. Taipei too is working on the construction and smooth operation of subways. Another major challenge is how to solve problems related to distribution services.

Whatever cities do, they will need vast sums in financing to build well-paved roads (especially expressways) and other infrastructure, not to mention the more preferable public transit systems. This will require governments to provide fiscal investments, as well as long-term financing with huge sums at low interest rates. Except for countries and regions like Singapore, Taiwan, and South Korea, however, Asian countries generally have little in the way of such funding. While all of them have high rates of economic growth and rising national income levels, it is difficult for them to domestically procure funds to build roads and public transit systems. For that reason they take the easy route by depending on foreign aid and foreign capital (BOT schemes), but these involve problems that need careful consideration.

Generally China, Indonesia, and other countries rely

heavily on customs duties or indirect taxes as the main sources of national tax revenues, thereby limiting the increase in overall revenues. They will perhaps find it necessary to establish and increase direct taxing mechanisms such as income tax, corporate tax, and inheritance tax (most Asian countries have no inheritance or gift taxes). They will also have to establish and improve systems for tax collection, accounting and bookkeeping, accounting for corporations, and the like. What is more, Asian countries all have insufficient financing on a long-term, low-interest basis. Their institutions that correspond to Japan's fiscal investment and loan program are weak. More fundamental is the necessity to increase their citizens' savings rates, which will in turn require increasing the citizens' trust in their own countries' financial systems and raising the citizens' consciousness with regard to savings. An example of problems in a non-Asian country would be Brazil a decade ago, where repeated changes in the currency system and extremely high inflation in the past resulted in very low trust by ordinary citizens in their own country's financial institutions and in savings, which was why high-income citizens send much of their money abroad.

Whatever the case, a major part of the solution involving national economies will be how to fairly and efficiently expand direct taxation, raise the citizens' savings rates, use those savings in a planned manner for providing roads, and, even more, to build and maintain public transit systems.

4. Motor Vehicle Pollution in Asian Countries

4-1 Short- and Long-Term Motor Vehicle Pollution Measures

Over the short term, measures to cope with motor vehicle pollution in Asia's rapidly developing large cities focus on these three items: (1) Fuel modification, (2) controls on individual vehicles, and (3) remedying traffic congestion. Fuel modification involves the big job of switching to unleaded gasoline, which makes it important to expedite investment for that purpose in oil refining. Cost/benefit assessments too show this to be one of the most effective ways to fight air pollution. It is also vital to reduce diesel fuel sulfur content. In Bangkok a citizens' organization called the Anti Air Pollution & Environmental Protection Foundation once helped the switch to unleaded gasoline by boycotting leaded gasoline, but economic incentives such as raising taxes on leaded gasoline should also be considered. Asian countries should also consider fuel controls to reduce benzene and other carcinogens in the hydrocarbons of motor vehicle exhaust.

Effective short-term ways to deal with vehicle pollution are controls on individual vehicles or strengthening emission controls on vehicles in use. With regard to new vehicles, Asian countries have learned from European and Japanese controls on individual vehicles in that they have in

recent years implemented strict exhaust controls involving unleaded gasoline and the use of catalytic converters and other technologies to reduce emissions. The problem is how to control used vehicles, and especially motorcycles with two-stroke engines. Apparently the World Bank and other institutions considered an option for totally eliminating motorcycles with two-stroke engines from Asian developing countries, and indeed all these countries are wondering how they can regulate used and poorly maintained motorcycles. In this respect they should perhaps give specific consideration to measures such as switching to four-stroke motorcycles, financial incentives for encouraging the switch to motorcycles that qualify under exhaust standards, barring disqualified motorcycles from entering urban centers, and designating electric bicycles as low-pollution motorbikes and subsidizing them. In Manila a joint campaign by the Natural Resources and Environment Ministry and citizens to control motor vehicle particulate emissions has been effective, a matter of particular interest because it shows the important role of participation by and increased environmental consciousness among the citizens in controlling vehicle pollution.

4-2 Managing Demand for Vehicle Use

Effective ways of dealing with traffic congestion include restricting traffic, establishing bus lanes, and measures to manage demand for vehicle use (parking fees, staggered working hours, ride-sharing schemes, etc.). A characteristic of problems encountered in controlling vehicle pollution in Asia is that these countries need the same kind of remedial measures as the developed countries with mature car societies even though they are still in the initial or median stages of building motorized societies. This is because traffic volume soars in these countries before they have built roads and developed their ancillary facilities and institutions, with the result being very serious traffic congestion in the early stage. An objective assessment of this situation shows that they must control demand for vehicle use in order to hold down traffic volume. Already the developed countries are working on policies to hold down motor vehicle use, and encourage the use of public transit and bicycles. In some places such policies are already being implemented. Some people argue that in order to contain excessive motor vehicle traffic in big cities, it is important that policies control urban growth through means such as checking population concentration in big cities and dispersing population, and that cities implement measures to control vehicle demand, such as ride-sharing systems, encouraging park-and-ride in the suburbs (for example, providing large parking lots near suburban train stations, having people park their cars and take trains into the city), building light rail transit (LRT) in city centers, limiting vehicle ownership, and using road pricing.

Of course some of these demand reduction measures are already being used in some large Asian cities. For example, beginning in 1992 Jakarta encouraged ride-sharing by allowing passenger cars with at least three riders into controlled city areas, while restricting those with only one or two occupants. Although ride-sharing schemes like this started in the U.S., they are also used in Indonesia, where private cars are still owned only by a small social elite. This creates another problem, however: Children are making money by getting picked up by drivers at the entrance to a controlled area and dropped off when leaving the area, thus allowing drivers to evade the restriction. In Jakarta even a new compact car costs 8,900 times an average daily wage (about 24 years' pay), and the same car even when used drops only to 3,800 times (about 10 years' pay), so even a generous estimate of households that can buy a car runs to a mere several percent. Meanwhile, Jakarta has poor public transit. Although bus fares must be officially authorized, they are not necessarily cheap (300-500 rupias, about 10% of daily wages). So a handful of car drivers, pedestrians breathing vehicle exhaust, riders of dangerous buses, and the like create a worsening situation of increasing transportation-related social inequality that will defy attempts at a solution.

4-3 Belated Infrastructure Development

Measures such as those described above are effective in the short term, and should be continued over the long term, but looking farther into the future, large cities will have to solve their transportation problems by building and enhancing mass transit systems that make it easier for the average citizen to move around the city. But essential requirements for this are plenty of funding and solid urban planning based on consideration of land use. As mentioned above, many developing countries presently get much of their funding for building roads and other infrastructure from multilateral or bilateral foreign aid. For example, Indonesia uses funds from abroad for about 40% of its investment outlays, with aid from Japan playing an especially big role. In fact, Japan's Overseas Economic Cooperation Fund (OECF) provides credit of 20 to 40 billion yen a year as "road network repair projects" and "regional road construction projects." The World Bank and Asian Development Bank provide the same kind of financing. Indonesia is the largest benefactor of foreign transportation-related financing by Japan, accounting for slightly less than 40% of the total.

Whether roads are built with foreign aid or BOT schemes, it will be especially important from now on that the environmental capacities of roads are planned in advance, and that roads are structured and land use planned to suit those capacities. In that respect Japan's experience has been unfortunate because, owing to the priority given to economic growth during the rapid-growth years, the ease of obtaining land led the country to build, in and around major cities, structures like two-tiered roads that lacked consideration for the environment, which resulted in serious noise, vibration, and air pollution along the roads. So as not to make the same mistake, it is important that other countries' road construction plans at the outset internalize the future social costs of motor vehicle pollution by restricting land use along roads,

widening rights-of-way to provide for green buffer zones, and take other measures when planning road construction. For those purposes they should always carry out comprehensive environmental impact assessments in advance and consider the best type of road structure (such as tunnels or semi-underground construction).

Meanwhile, rail transit plays almost no part in the urban transportation of Asia's big cities. In fact, if we take the example of Jakarta, the city has one of the finest rail systems among Asia's developing countries, but it plays an almost negligible role in urban transportation. Likewise in Bangkok, rail plays hardly any part at all, so it is buses that serve as public transportation. In view of this situation, the construction of public rail transit has become a major issue in Bangkok during recent years.

Bangkok is working on three rail construction plans for urban transportation. One is an elevated railway of about 20 km by the Metropolitan Rapid Transit Authority (MRTA). The government has already decided to put 11.3 km of this underground. In 1992 MRTA was separated from the Thai Interior Ministry's Expressway and Rapid Transit Authority (ETA) and became a government organization directly under the Prime Minister's Office for building and running high-speed rail transit in the Bangkok area. At first this plan was to be run directly by the government, but construction was switched to the private sector, where it remains today. Second is the plan for a Sky Train of 60.19 km built by Hopewell, a private-sector Hong Kong company. The plan calls for building roads and railways over existing State Railway of Thailand (SRT) tracks, and it is already about 40% complete. While this is a typical BOT scheme, it involves complicated problems including the connection with the Thai Ministry of Railroads and reconciliation with Hopewell's development right, so construction is not necessarily proceeding smoothly. Recently in Thailand land prices have plummeted owing to bursting of the real estate bubble, and construction costs have spiraled because of the baht's sharp decline, leading the Thai government to a decision to cancel its contract with Hopewell. Third is a plan by the Bangkok Metropolitan Administration (BMA) to build a 23.7 km electric railway. This too is a plan to build an elevated railway above an existing road, and is a typical BOT scheme. All these plans depend on private-sector capital and are closely watched to see how well they proceed and whether they succeed or fail.

5. Conclusion: Petroleum Consumption and Global Warming

Motor vehicle fuel accounts for a very high proportion of petroleum consumption in Asian developing countries. As a matter of fact, while the transportation sectors in the U.S. and Japan account for 27% and 23%, respectively, of total energy consumption, this proportion in Sri Lanka, Thailand, Malaysia, the Philippines, Nepal, Indonesia, and other countries runs from 61% to 33%. Thus in Asian countries

Fig. 5 Past and Projected Future Asian Proportion of World CO_2 Emissions

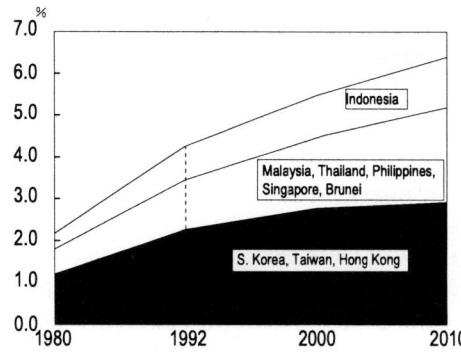

Source: Advisory Committee on International Trade and Industry, *Energy Vision for Asia*, 1995, ed. Ministry of International Trade and Industry, Agency of Natural Resources and Energy.

motor vehicle transportation makes an unusually large contribution to global warming. The proportion of worldwide CO_2 emissions by nine Asian countries and regions excluding China was over 2% in 1980, but is anticipated to exceed 6% in 2010 (Fig. 5), making it possible that new limitations will be imposed on increasing motor vehicle use in Asia. Further, the oil import bills of non-oil-producing developing countries account for an unusually high proportion of their total monetary export amounts; the two-thirds proportion for Bangladesh and one-third for South Korea and Thailand in the early 1980s just after the second oil crisis show that payments for imported oil were already an onerous burden. It is thus quite foreseeable that the possibilities for such petroleum imports in the future could be a characteristic limiting factor on the spread of motor vehicles in Asia.

In terms of future vehicle fleet growth, oil consumption is expected to skyrocket especially in China. As of 1995 China's fleet was 10.4 million vehicles, with the breakdown being nearly 8.3 million gasoline-powered vehicles and 2.1 million diesel-powered vehicles (35.0% of trucks are diesel), which consumed about 22.6 million tons of gasoline and 7.5 million tons of diesel fuel. Additionally, a number of predictions say that China's fleet will swell to about 18 million vehicles in 2000 and 40 million in 2010. If China's fleet grows this quickly, and even assuming that per-vehicle fuel consumption is 30% under that of 1994, in 2010 China's gasoline and diesel fuel consumption will increase by 2.6 and 4.9 times, respectively. In that year, China's oil consumption will be 1.4 times that used by Japan's fleet in 1994, which would make China a mega-consumer of oil products to run its vehicle fleet. These forecasts will change depending on how much China's currently very poor per-vehicle fuel economy improves, but whatever happens, we can see that as China's motor vehicle use increases, the country will consume more oil and emit more CO_2.

As a means of dealing with CO_2-induced global warming, developed countries in recent years have been seriously discussing the development and use of low-emission vehicles such as hybrids and natural gas vehicles, or zero-emission vehicles like electric vehicles. There is no doubt that the developed countries are technically capable of reducing CO_2 emissions to an extent by developing and promoting LEVs, but what about Asian developing countries like China? If we are contemplating real solutions for the present serious traffic congestion and the variety of pollution and environmental problems arising from motor vehicle use, a more pressing and important challenge than the development of LEVs is how these countries can build efficient public transit systems.

(NAGAI Susumu, SHIBATA Tokue, MIZUTANI Yoichi)

Essay   The Hazards of Airborne Particulates

The most serious component of the motor vehicle air pollution common to Asia's megacities is particulates, as typified by Bangkok. However, Asian countries do not all measure such air pollution with the same methods, and they have varying environmental standards, so comparisons between countries require special care. In fact, particulate measurement methods include (1) total suspended particulates (TSP), which measure particles regardless of size, (2) the Japanese method of measuring particles under 10 μm in size and disregarding all larger particles, and (3) PM_{10}, which disregards 50% of particles 10 μm or larger. Particulate sources differ depending on the measurement methods used. In developing Asian countries the sources are not only diesel exhaust particulates (DEPs), but also other anthropogenic sources such as factories, construction work, and road dust, as well as soil particles and other natural sources. Thus particulate control measures include (1) preventing dust emissions from factories, (2) using tarps at construction sites, (3) restricting open burning, and (4) sprinkling water on streets.

The most hazardous particulates at present are said to be fine particles under 2 μm in size. Because of their small size they cause serious health damage by penetrating the deepest recesses of human lungs and crippling lung functions. Fine DEPs in particular contain many chemical compounds, some of which are carcinogenic or mutagenic. The International Agency for Research on Cancer has assessed benzo(a)pyrene and benzo(k)fluoranthene as carcinogens, and it is known that their concentrations in particulates are higher in fine particles.

In view of these facts, we cannot correctly assess the hazards of airborne particulates merely by simply comparing their atmospheric concentrations. There are variations in the standards set by each country. For example, while the WHO guideline is 150 μg/m^3 for the 24-hour TSP value, Japan's is 100 μg/m^3 for airborne particulates. As noted in Chapter 2 above, in Thailand it is 330 μg/m^3 for the 24-hour TSP value (Thailand has other measures such as a yearly average of 100 μg/m^3), and Indonesia has standards including 260 μg/m^3 for the 24-hour TSP value. The hazards of these airborne particulates must be assessed in accordance with the proportion from motor vehicle exhaust. In Thailand, for instance, the hazards are especially great because the proportion of diesel vehicles is high. The reason for this preponderance and the unusually high 60% proportion for diesel fuel in total motor vehicle fuel consumption is that small pickup trucks are taxed lower than passenger cars and therefore used in their place, which is a major source of DEPs.

(NAGAI Susumu)

Chapter 3
Pollution and Health Damage

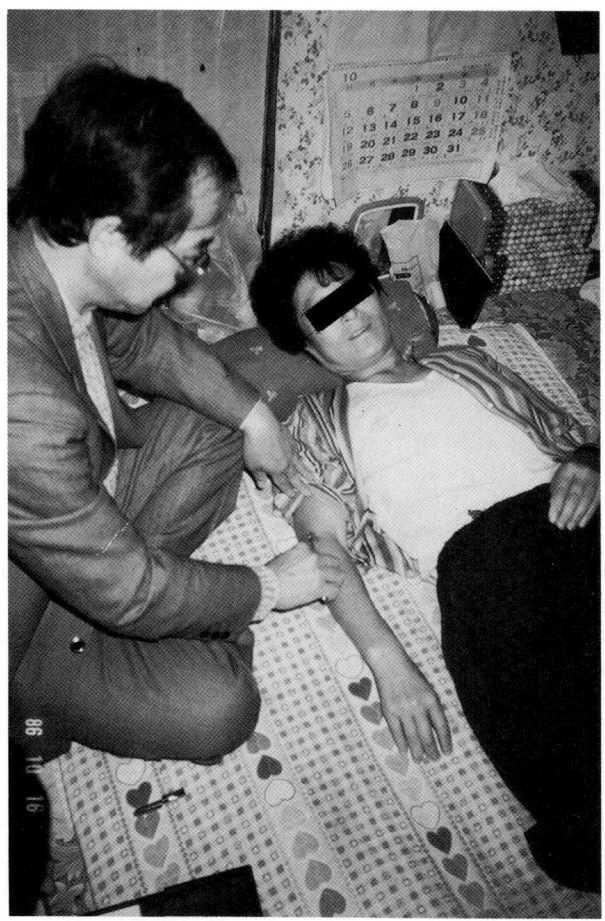

Harada Masazumi, one of this chapter's authors, examines a pollution victim near the Onsan industrial complex. Photo courtesy of author.

1. Introduction: Compound and Diverse Pollution

Asian countries are highly diverse in their politics, economies, cultures, and histories, as well as the natural environments that shape them. But just as diverse are the many kinds of pollution, their effects on human health, and their environmental damage.

Since the 1992 Rio Earth Summit Japan's government has pledged to make international contributions toward solving environmental problems, especially in Asia. Heretofore Japan has indeed accumulated a measure of expertise that would, in the technical and legal senses, qualify it as a country advanced in dealing with pollution, a qualification earned while causing serious pollution that damaged the health of many victims, and then paying for it with their blood and tears. So if Japan makes positive use of the know-how yielded by those bitter experiences, it can to an extent perhaps contribute to alleviating the many kinds of pollution now occurring across Asia. Nevertheless, though in some ways the problems in other Asian countries are the same as what Japan once had, in many ways they differ. Japan's experiences and know-how cannot necessarily be applied as is to solve the problems of other Asian countries.

At the same time, it is an indisputable fact that Japan's economic development has become a sort of model to Asian developing countries. Japan is given a place in world history as having achieved "miraculous economic development," but Japan started the journey toward so-called modernization and industrialization well over a century ago in the late 1860s. From the perspective of industrial structure, this development began with mining and its associated industries, then progressively to steelmaking, heavy and chemical industries, pulp manufacturing, the electrical industry, and chemicals, then to petrochemicals and high technology. In conjunction with that progression, the types of pollution and their principal causes changed. For example, instances of pollution that arose from mining were the soil, water, and air pollution at Ashio, Besshi, Hitachi, and other places, cadmium poisoning that caused itai-itai disease, and arsenic poisoning at Toroku in Miyazaki Prefecture. At Minamata the cause was an electrosynthesis chemical factory that emitted methylmercury. The cause of Yokkaichi asthma was pollution from a number of petroleum-related industries, mainly SOx. In Osaka's Nishiyodogawa area expressway motor vehicles were the source of pollution, mainly SOx and NOx. Until now, pollution in Japan has been caused by multiple pollutants in some cases, but mainly by single pollutants, and in many cases the offending companies could be identified. Fortunately, this perhaps made it comparatively easy to determine causes and prove causality epidemiologically, but this is not necessarily so for the problems other Asian countries now face. In fact, their pollution from the outset has basically been compound, caused by the simultaneous spread of a large variety of pollutants.[1]

---

[1] For information to supplement this section, see Ui Jun, ed., *Industrial Pollution in Japan*, UN University Press, Tokyo, 1992.

2. Pollution Diseases in Developing Asian Countries

Asian developing countries are trying to achieve in a mere one or two decades the modernization and industrialization that Japan took over a century to accomplish, resulting in the simultaneous appearance of nearly all kinds of industries, from mining to modern hi-tech. It would be as if Japan's Minamata were home to the industrial complexes of Yokkaichi, Tokuyama, and Omuta, with the mines of Ashio added. So these other countries have a mix of pollutants from the very outset, and on a broad scale.

An additional problem is that large-scale development processes are on a new dimension that magnifies preexisting classical pollution problems. With this situation in mind, we can point out at least the following four characteristics of principally industrial pollution in Asian developing countries.

(1) Compound pollution: The rapid development and simultaneous operation of new and existing industries of various kinds cause compound pollution via many pathways (soil, water, air, food, etc.).

(2) Extensive pollution: Rapid and historically unprecedented large-scale development brings about widespread pollution of a kind not experienced in Japan.

(3) Internationalization: Technology transfer, direct investment, and other actions by the developed nations constitute one factor underlying pollution and environmental damage. In our modern age, international political and economic relationships in particular are much closer, and information is available at real time, so countries strongly influence each other. But there is a quickly growing imbalance in that while physical distribution grows to huge proportions and technologies are transferred ever faster, management in terms of the environment, public health, safety, and other concerns is totally unable to keep pace, thereby making these countries prone to broad-scale disasters and pollution.

(4) Invisible health damage: In addition to compound pollution and new kinds of pollution, overall poor public health management creates the potential for pollution-caused health damage that is hard to perceive.

Below we shall examine several specific examples that illustrate the characteristics of pollution problems in Asian countries.

3. Multiple Pollutant Health Damage: Onsan, South Korea

A typical example of compound pollution would be that near the Onsan industrial complex in South Korea. As Part II Chapter 2 discusses in detail, since 1974 when the South Korean government designated the Onsan area for development of an industrial complex, this approximately 2,000-ha area has been crowded with industries including the refining of lead, copper, zinc, aluminum, and other nonferrous

Table 1 Onsan Disease (South Korea)

Figures are percentages

Symptoms	Eye	Respiratory	Digestive	Skin	Nerves
Seoul National University (1984)	72.2	48.2	22.2	59.4	34.2
Environment Agency (1985)	11.9	10.3	5.5	9.0	53.9
Harada (1986)	50	40	30	20	75

Source: Harada Masazumi, "Environmental Sciences", 4, Suppl., S157(1996).

metals, with textiles, pulp, chemicals, fertilizer, dyes, and petroleum refining and other petrochemical plants, as well as with hi-tech industries that have caused serious compound pollution of the air, soil, and water. Actual operations began in 1978 by Koryo Zinc Co., Ltd. (a joint venture with Japan's Toho Zinc Co., Ltd.), and already the next November there was an electrolyte leak from a chemical plant that caused serious damage to farmed seaweed. In 1982 a gas leak from another chemical plant resulted in the hospitalization of 105 local citizens. In 1984 there was another gas leak accident because of which nine children in a nearby schoolyard collapsed and were hospitalized. Then in 1985 the South Korean media caused a furor by reporting that 500 Onsan residents were suffering from itai-itai disease (which is characterized by kidney damage, osteomalacia, etc.). It turned out not to be itai-itai disease, but subsequent studies by Seoul National University, South Korea's Environment Agency, and Harada Masazumi confirmed that local citizens were indeed suffering from various kinds of health damage (Table 1). Among the residents the studies found high incidences of symptoms in their eyes, respiratory systems, digestive systems, skin, and nerves (including muscles and joints). In particular, all victims had pain in their muscles, joints, and nerves. Noteworthy are the facts that symptoms were manifested in many cases regardless of age, and that they appeared after 1980 when pollution became serious. Studies also found geographical differences in prevailing symptoms depending on the kinds of factories near people's homes. In some places respiratory, eye, and cutaneous symptoms prevailed, while in others nerve (primarily peripheral) symptoms were salient. These differences likely indicate causality between symptoms and the factories.

There are, however, almost no other instances of such health damage by compound pollution, making it very hard to determine which substances are at fault, and to prove causality. Perhaps the only name for this is indeed "Onsan disease." At first South Korea's Environment Agency did not recognize this as a pollution-induced illness, but it did acknowledge the heavy pollution itself, and recommended that affected people relocate. Nevertheless, factories continued to proliferate and numbered over 120 after 1990, employing nearly 10,000 workers (Table 2).

Owing to these events, about 30,000 people abandoned their homes here beginning in 1991 and relocated before the extent of damage could be determined. According to witnesses, some of these people found it impossible to maintain their livelihoods in their new domiciles, returned to Onsan in secret, and made a living by poaching fish. And even if relocated citizens' problems are considered solved by moving, there are still concerns about health damage to the many workers employed in Onsan.

Further, near South Korea's Yosan industrial complex, which similarly has a concentration of 97 facilities including large petrochemical plants, steelmaking plants, and coal-fired power stations, there are many cases of respiratory and skin disorders. The same thing could very well happen in developing countries with similar plans for industrial complexes. Indeed, problems have already appeared at the Philippines' Leyte industrial complex, Malaysia's Pry industrial complex, and Thailand's Lamphun industrial complex, but unfortunately there are still insufficient clinical epidemiological studies of these areas. Other important features are that in almost every case these large-scale industrial complexes are part of national policy, and that they host many highly polluting transnational corporations. Indeed, 75% of the joint ventures at Onsan have Japanese partners. These situations make it very difficult even to elucidate the extent of pollution and health damage.

4. Widespread Groundwater Contamination

A discussion of the second characteristic cannot go without mention of the widespread occurrences throughout Asia of poisoning, a classic problem in the developed countries. A typical and especially grave example of this would be many victims poisoned by arsenic contamination (Fig. 1, Table 3).

For example, the arsenic contamination that occurred in West Bengal (India) and Bangladesh, on the Ganges River, is said to have exposed 20 million people. Apparently

Table 2 Number of Plants in the Onsan Industrial Complex

	1984	1990	After 1990
Textiles			3
Wood products			2
Pulp		1	2
Petrochemicals	7	29	68
Nonmetals		2	3
Primary metals	5	7	12
Prefabricated metals		11	32
Others	1	2	2
Total	13	52	124
Workers	5,508	8,111	9,276

Source: Same as Table 1.

Fig. 1 Arsenic Poisoning Sites in Asia

Source: Harada Masazumi, *Digest of Science of Labor* (in Japanese), vol. 52, no. 3 (1997).

220,000 people in India alone fell victim to arsenic poisoning. There is a close connection between this and the heavy use of groundwater to increase food production and accommodate population growth. Groundwater use was necessary because dam construction and logging of forests had depleted supplies of irrigation water, and this situation was further hastened by the economical and technological ease of digging deep wells. Pumping groundwater in large quantities allowed a great deal of arsenic to leach out of the ground, making irrigation directly responsible for the poisoning. In India's West Bengal, for example, an average

0.25 ppm, and maximum 3.7 ppm, of arsenic was detected in drinking water, while the safety standard is 0.05 ppm.

Arsenic poisoning itself has been known since long ago. Its main symptoms are melanosis, hyperkeratosis, leukoderma, and other skin symptoms. It also affects the blood vessels, liver, heart, and other organs, and causes cancer after 10 or 20 years. Serious arsenic groundwater contamination like this is appearing in a number of places throughout Asia. In addition to West Bengal, there is confirmation of many more victims including 13,000 in southern Taiwan, 1,600 in Inner Mongolia, and 2,000 in China's Xinjiang Uighur region. There are no doubt many more as yet undiscovered.

At Ron Phibun in southern Thailand arsenic contamination of groundwater by tin mining resulted in a reported 1,400 victims of poisoning. Tin ore in this area of southern Thailand contains 53.6 ppm arsenic as an impurity, and tests detected a maximum concentration of 5,209 ppm in soil, and 4.45 ppm in drinking water. The presence of many operating and abandoned tin mines extending from this area into Malaysia makes it possible that more widespread groundwater contamination by arsenic will be discovered in the future.

The foregoing examples are perhaps typical of groundwater contamination by arsenic, but the trend in developing countries is for widespread occurrences of pollution that is not limited to groundwater contamination. For example, the mercury used in gold mining is released into the environment where it methylates, accumulates in fish and shellfish, and harms people who eat these organisms. This kind of harm (Minamata disease) is likewise characterized by the large areas and populations affected, and is already a serious problem in Brazil's Amazon basin and in the Philippines.

Table 3 Arsenic Poisoning in Asia

Country/location	Year discovered	No. of victims	Cause	Highest ppm	Notes
Japan/Nakajo (Niigata)	1954	93	Groundwater contamination by arsenic factory	4.0	Studied again 28 years later; high incidence of lung cancer.
Japan/Toroku (Miyazaki)	1971	144	Arsenious acid calcination	1.07 in water	Multiple pathways (food, air, water), and serious soil contamination.
China/Guizhou	1953	3,000	Coal combustion	100-9,600 in coal	*Laizi* disease, an endemic disease; multiple pathways.
Taiwan/Tainan	1956	13,000	Groundwater	1.82	Black foot disease (gangrene), an endemic disease.
China/Kuitun	1980	2,000	Groundwater	0.85	Compound contamination with fluorine.
India/West Bengal	1983	220,000	Groundwater	3.7	Groundwater pumping for irrigation.
Thailand/Ron Phibun	1987	1,400	Groundwater contamination by tin mine	4.45	*Wujiaobing* (black fever), an endemic disease.
Inner Mongolia	1988	1,600	Groundwater	1.86	Fluorine poisoning; bone disorders 230,000 people, teeth disorders 1.9 million.
Vietnam/Son La	1990	166	Mining	1.14 in water	Multiple pathways (air, food, water).

Source: Harada Masazumi, *Environmental Science* (in English), vol. 4, suppl. S157-169 (1996), with some modifications. Table prepared by the Asia Arsenic Network.
Harada Masazumi, *Digest of Science of Labor* (in Japanese), vol. 52, no. 3 (1997).

5. Internationalization of Problems: Involvement of Developed Countries

Let us proceed to examples illustrating the third characteristic. Already well known are incidents of mercury contamination that occurred in the latter half of the 1970s in Thailand's Chao Phraya River watershed and Indonesia's Jakarta Bay. Fortunately, both were resolved without any health damage, but the sources of pollution were thought to be caustic soda plants that had come from Japan. At that time in Japan the manufacture of caustic soda was switching to other methods not requiring the use of mercury as a catalyst owing to the Minamata disease incident, so these overseas factories led to claims that Japan was exporting its pollution to other countries. As noted in connection with Onsan, large industrial complexes in developing countries generally include corporations from developed countries, and not a few of them are polluting industries. Following are a few representative examples.

(1) On December 3, 1984 at Bhopal, a city in central India, a pesticide plant of the U.S. company Union Carbide accidentally released methyl isocyanate (MIC) gas, causing the deaths of 2,000 and the serious poisoning of about 50,000 people. MIC is an intermediate product in the manufacture of carbamate pesticides that in developed countries is subject to strict storage controls, yet 45 tons of it were stored in Bhopal. Until this major accident there had reportedly been frequent small leaks, and although the plant had been equipped with five safety systems, all of them failed to operate on the day of the accident, strongly suggesting errors in safety management. Even the victims fortunate enough to survive are afflicted with respiratory disorders, eye symptoms, digestive disorders, nervous symptoms, and other grave aftereffects (Table 4). Reports say that as of six years after the accident 4,500 people have died.

(2) At Bukit Merah Village in the central Malaysian state of Perax, the Japanese-capitalized company Asian Rare Earth Sdn. Bhd. began the mining and refining of rare earth metals in 1982. The yttrium, scandium, lanthanum, cerium, and other rare earths obtained there were all exported to Japan, but a byproduct of their refining was the radioactive element thorium 232 (7% by weight). With a half-life of 14 billion years, thorium 232 demands strict control, lack of which resulted in contamination of the surrounding environment. Health damage included many cases of leukemia among the children living in the area, and women who had been employed at ARE gave birth to boys with microcephaly (lowered intelligence) and congenital cataracts, prompting suspicions that the plant environment was to blame. ARE has been shut down since 1992 by a lawsuit (see Part II Chapter 4 for details).

(3) The Lamphun Industrial Complex in northern Thailand's Lamphun State near the city of Chiang Mai hosts 86 companies, and 16 of the 20 electrical component companies are from Japan. Problems at this industrial park are serious pollution and occupational illnesses, but there is still little understanding of the situation and the causes. Assuming multiple pollutants, it will not be easy to demonstrate

Table 4 Clinical Symptoms of MIC Gas Exposure (70 Cases)
(Figures are percentages)

Symptoms	Immediately after exposure	One year later
Eyes	97.1	80.0
Pain / burning	92.9	65.7
Hyperemia	90.0	12.9
Lacrimation	87.1	51.4
Impaired eyesight	64.3	60.0
Respiratory symptoms	95.7	88.2
Dyspnea	85.7	78.6
Chest pains	75.7	48.6
Coughing	88.6	48.6
Phlegm	67.1	25.7
Sore throat	72.9	18.6
Asthma	-	14.3
Digestive symptoms	81.1	77.1
Nausea	52.9	25.7
Vomiting	51.4	10.0
Abdominal pains	60.0	31.4
Diarrhea	30.0	7.1
Anorexia	61.4	61.4
Loss of consciousness	34.3	1.4
Headache	78.6	70.0
Muscular weakness	72.9	64.3
Muscle pains	25.7	31.4
Loss of memory	11.4	20.0
Depression	0.2	0.8
Dermatitis	12.8	8.6
Eczema	1.4	0.0
Urticaria	5.7	7.1
Causalgia	22.7	12.9
Abnormal vaginal discharge	57.1	57.1
Irregular menstruation	21.4	12.9

Source: Harada Masazumi, *Digest of Science of Labor* (in Japanese), vol. 41, no. 6 (1986).

causality, but epidemiological studies are urgently needed.

(4) On May 10, 1993 a fire broke out at the Kader toy factory on the outskirts of Bangkok, with toxic gas killing 188 and injuring 400–500. As this was a foreign-capital company, the incident developed into a social issue.

The foregoing incidents might be called typical examples of pollution export. Behind the problems created by instances of pollution export like these one will find an elaborate weave of international political and economic interconnections, which requires host countries to consider new measures, differing from those for addressing purely domestic pollution problems, in order to deal with pollution sources and accidents, and to compensate for damage.

6. Invisible Pollution Diseases

The fourth characteristic of pollution problems in Asia, especially developing countries, is that health damage proceeds unseen as what might be called "invisible" or "hard-to-see" pollution diseases.

Normally, making pollution-caused health damage apparent requires improvements in public health and the other overall conditions under which we live. An example that demonstrates this point is the Ashio mine poisoning incident in prewar Japan. Despite the appalling environmental damage and pollution that mining caused, contemporary records make hardly any mention at all of human health damage because in those days much of the health damage caused by serious pollution was totally overlooked, masked as it was by problems such as tuberculosis, parasites, and malnutrition. As early as the 1910s there were apparently already victims of itai-itai disease, but the malady was not officially discovered until 1957. Likewise, pollution at Toroku occurred before the war, but it was 1973 before it was at last officially recognized. This is likely happening in other Asian countries now.

For example, in 1983 people in Japan heard about the appearance of congenital or infantile Minamata disease among fisherfolk along the shore of Indonesia's Jakarta Bay. Immediate on-site investigation found that already 28 infants and small children had died of a central nervous system disorder with unknown cause, and six were still living. Their symptoms indeed resembled those of Minamata disease, but low mercury values in their scalp hair and preserved umbilical cords showed that it was some other, unknown cause. At that time Indonesia's mortality rate for children up to five years old was 11.1%, or 4 to 5 times higher than Japan's, which suggested some kind of infection. But there is no doubt that gold miners in the Philippines and deep in Brazil's Amazon basin were suffering from inorganic mercury poisoning. Further, scalp hair from fisherfolk downstream on the Amazon showed high mercury levels, some of them already exceeding safety standards, but in fact a far greater number of deaths in these places are due to malaria and cholera than to inorganic mercury poisoning or Minamata disease. Thus pollution-caused health damage is very hard to see because it must vie for attention with malnutrition, infections, and other problems.

Additional factors are that in situations where there are multiple pollutants, as noted in the cases of South Korea's Onsan and Thailand's Lamphun, demonstrating which substances are the cause is extremely difficult; moreover, when a cause is small amounts of pollution over a long term, determining the existence of health damage itself is not easy. For example, confirmation of Minamata disease caused by mercury pollution from the Jilin industrial complex in northern China took a long time owing to the comparatively minor symptoms. Detecting unusual health patterns in people exposed to compound pollution or long-term, low-concentration pollution therefore makes it essential to conduct persistent and patient epidemiological studies while comparing with control groups in unpolluted areas. In Japan there was protracted debate over the definitions of even Minamata disease and itai-itai disease, which are very definite pollution-induced types of health

damage. These bitter experiences in Japan should serve as a lesson for the developing countries.

Furthermore, it is hard to demonstrate causality for health damage that is totally without historical precedent, as with the thorium 232 contamination in Malaysia's Bukit Merah Village and the MIC poisoning in Bhopal. Similarly, in cases like the dioxin contamination caused by defoliants used in the Vietnam war, it is not clear even what health effects will occur because such a thing had never happened before. Only in recent years has this problem come to the fore, mainly in the developed countries, as a new challenge involving controls on chemical substances of questionable safety and on hazardous wastes. But many developing countries where controls are nonexistent or very weak will need to quickly implement appropriate measures to ensure that they do not become human experimentation laboratories for the effects of such hazardous substances.

7. Summation: How Useful Are Japan's Experiences and Lessons?

As mentioned at the outset, Asia is not a homogeneous monolith. Similarly, Asian countries face a wide variety of pollutants and their resulting health damage, so it is unlikely that Japan's experiences and the lessons learned would be useful to other countries as is. Unfortunately, if the present situation continues, economic development in Asia will bring about pollution and the attendant health damage, which will predictably worsen. Serious thought is needed about what Japan can and should do. It is very important that Japan not simply consider what kinds of aid are appropriate for correcting economic disparities between developed and developing countries, and for alleviating poverty, improving public health, or solving environmental problems, but also to question our own excessive consumption, which would not be possible without plundering Asia's environment and resources, and to give other countries an accurate account of the down side of Japan's experience and bitter lessons learned heretofore. In this respect, air pollution by the Lampang thermal power plant in northern Thailand is of particular interest. The local citizens' movement, having learned from Japan's pollution experience, sought health examinations to elucidate the state of their own health (Table 5), and obtained compensation and health care for pollution victims. They were also successful in having air monitoring stations installed in 15 locations in local villages, and concluding an agreement to have the plant curtail operations when pollution exceeded standards. Asia will need intergovernmental exchanges and aid, but in addition it will be increasingly important to have citizen-level interchanges and the development of international networks based on such interchanges.

Information and data of all kinds from Asian countries are available today with far greater ease than before, but for some reason concrete information and data on pollution and its resulting health damage are very hard to come by. One possible reason is that Asian countries simply have not been

Table 5 Air Pollution Health Damage from the Lampang Power Plant in Thailand (1992)

Persons examined	3,237
Respiratory infections	3,031
Digestive disorders	505
Muscle and joint pains	735
Skin symptoms	203
Eye symptoms	203
Headaches	291
Neurasthenia	193
Gingivitis, stomatitis	89
Dizziness, nausea	76

Source: Electric General Authority Thailand (EGAT), 1992.

able to arrange for field studies. In this respect there is certainly a great deal of opportunity for us to build international cooperation with the citizens and experts of other Asian countries in areas such as field studies and policy research relating to environmental problems including health damage, where Japan has accumulated a great deal of knowledge and expertise from its own pollution experience.[2]

(HARADA Masazumi, TERANISHI Shun'ichi)

---

[2]   Harada, M. "Characteristics of Industrial Poisoning and Environmental Contamination in Developing Countries," *Environ. Sciences*, 4; Suppl. S157, 1996.

Essay   The Original Pollution Disease: Lessons of Minamata Disease[3]

Discovery of Minamata disease On May 1, 1956 physicians at the Chisso Corporation hospital contacted the Minamata Public Health Center and reported that there were many cases of a central nervous system disorder with unknown cause among fisherfolk in the coastal areas of Minamata City. This was the official discovery of Minamata disease. The disease was first discovered by its prevalence among children, which indicates that the first people in a given area to suffer health damage from pollution are the physiologically weak, such as small children, the unborn, the aged, and the ill. An infection (encephalitis) was suspected at first, but immediately found not to be the case. Immediately the medical association, municipal hospital, Chisso Corporation hospital, Public Health Center, and other institutions began to investigate the cause, discovering that already in 1954 there had been victims including adults. That August the Kumamoto University School of Medicine organized a research team to find the cause, ascertaining that victims were localized in rural areas along the coast of Minamata Bay, with fishing families prevalent and no connection to age or sex, that despite close living conditions of families there was no chain contagion trend, that victims ate many fish and shellfish from Minamata Bay, and that in areas with many victims there were also many cat deaths with the same symptoms. Minamata disease was therefore clearly some kind of poisoning. Such being the case, the Chisso Minamata chemical plant on the bay was suspect from the outset because its effluent discharge port opened into the bay. Yet, both Chisso and the local government did nothing, claiming that the cause was unknown.

Unusual occurrences in nature presage pollution Compensation had already been paid several times since 1925 for fishing damage caused by Chisso Minamata plant effluent. In such cases the most important thing is stopping the pollution, not paying compensation, but Chisso officials publicly stated that because the chemical industry is essential to Japan's economic development, a certain amount of damage to fisheries cannot be helped. Chisso solved all problems with money and did nothing to stop the pollution. Around 1950 unusual phenomena became noticeable in the environment. Minamata Bay octopuses, sea bass, and the like floated to the surface where they could be caught by hand, and people observed that oysters near Chisso's wash port were dead. In 1953 cats began dying in unusual ways, and the following year cats disappeared from fishing areas of Minamata City. People saw cats drooling, wavering, and staggering as they walked; cats would suddenly spin and jump violently with convulsions, or charge into the sea. People were disturbed by the eeriness, calling it cat suicides or cat dancing sickness. Subsequently fish were found floating, waterfowl fell from the sky, and animals including pigs, dogs, and chickens died in mad frenzies. Unusual occurrences in nature presage pollution-caused human health damage. Such was the start of the biggest pollution incident in human history.

Search for the poison The Kumamoto University research team knew nothing about how the Minamata plant operated, and Chisso did not cooperate, so finding the cause of Minamata disease was difficult. In 1958 researchers determined that the clinical symptoms shared by victims included constriction of the visual field, sensory disturbance, disturbance of muscular coordination, disarthria, hearing impairment, and tremors. Pathological characteristics revealed by autopsies were specific damage to sensory and

---

[3]   Harada, M. "Minamata Disease: Methylmercury Poisoning in Japan Caused by Environmental Pollution," *Critical Reviews in Toxicology*, 25(1): 1-24 (1995

motor centers, the hearing center, and the vision center in the cerebral cortex; and the granular cell layer in the cerebellum. The only cases that matched illnesses with these characteristics were the methylmercury poisoning cases of pesticide plant workers in England that were made public in 1940, for which reason the research team immediately began to check for mercury. An analysis of mud near the plant's effluent port by Prof. Kitamura Masatsugu and others detected 2,010 ppm mercury (wet weight), and mercury was found in high concentrations in mud throughout the bay. Bay shellfish had a maximum concentration of 39.0 ppm. High concentrations were also detected in finfish and other organisms, with maximum values such as 14.9 ppm in white croaker, 16.6 in sea bass, 24.1 in sea bream, and 35.7 in crabs. Because fish move about, contaminated fish were discovered throughout the Shiranui Sea. Cats with naturally occurring Minamata disease had maximum values of 101.1 ppm mercury in their livers and 52.0 in their fur. Maximum mercury concentrations in cats fed Minamata Bay fish and shellfish to induce symptoms were 145.5 ppm in the liver, 70.0 in the fur, and 18.6 in the brain. Cats brought from Kumamoto City to live in areas with high incidences of Minamata disease all developed symptoms in 33 to 65 days. Internal organs of deceased Minamata disease victims also had high mercury concentrations. Kitamura began measuring mercury in scalp hair as an indicator of bodily contamination. As sampling and measuring mercury is easy with scalp hair, and because it reveals the total contamination for a certain period of time, the method is used the world over and is very useful in diagnosis and prevention. Minamata disease victims had a maximum scalp hair mercury value of 705 ppm, while a maximum value of 191.0 was found in the scalp hair of Minamata residents and victims' family members who exhibited no symptoms.

Minamata disease found to be organic (methyl) mercury poisoning On July 14, 1959 Kumamoto University announced that Minamata disease was organic mercury poisoning. Grounds for this included: clinical symptoms and pathological findings matched organic (methyl) mercury poisoning; mercury had been detected in high concentrations; and cats with naturally occurring Minamata disease were identical both symptomatically and pathologically with cats fed fish and shellfish, and with cats administered methylmercury. In February 1960 Uchida Makio succeeded in extracting and crystallizing methylmercury compounds from Minamata Bay short-necked clams, which made methylmercury poisoning definite. In October 1961 Irukayama Katsuro extracted methylmercury chloride from the sludge produced by the acetaldehyde process, which proved that methylation was occurring inside Chisso's plant. This scientific sleuthing demonstrated the cause of Minamata disease, which occurred through unprecedentedly widespread environmental contamination and through the food chain.

Congenital Minamata disease Researchers noticed that from the beginning many people with cerebral paralysis were born in areas with high incidences of Minamata disease, but they had not eaten contaminated fish. At that time people believed that toxins did not pass through the placenta. Between 100% and 75% of such patients shared these symptoms: mental retardation, disarthria, hyperkinesis, ataxia, deformities of limbs, strabismus, hypersalivation, spastic symptoms, abnormal reflexes, and impaired growth and nutrition. Furthermore, these appeared at a 9% rate in the fishing villages with the highest incidences of Minamata disease, and their occurrence coincided both with the acute appearance of Minamata disease both in terms of geography and time period. Acute Minamata disease was found in 64% of their families, and later chronic Minamata disease was found in 100%. Victims' mothers frequently ate fish, and exhibited symptoms of sensory disturbance and, although minor, constriction of the visual field and ataxia. In August 1962 a congenital Minamata disease victim died and was autopsied, pathologically revealing that the cause was methylmercury poisoning while still in the womb, and making known for the first time in the world the existence of congenital Minamata disease. Sixty-four cases have been confirmed to date, 13 of whom have died. Researchers subsequently confirmed experimentally that methylmercury passes through the placenta.

Chronic Minamata disease In 1965 the second outbreak of Minamata disease was discovered on the lower reaches of the Agano River in Niigata Prefecture. It too was caused by acetaldehyde manufacture, this time by the Showa Denko plant 60 km upstream, which contaminated fish and shellfish with methylmercury, thereby poisoning victims. Citizens in the most heavily contaminated area were examined and their hair tested for mercury, showing that the definition of Minamata disease until that time had been too narrow, and revealing the existence of chronic and delayed-onset Minamata disease. In June 1967 victims filed a civil lawsuit in Niigata charging the company with liability, which led to a review of the disease definition in Minamata (i.e., the search for other potential victims) as well, and the filing of a lawsuit against Chisso in June 1969 seeking payment of compensation. Victories for plaintiffs (victims) in both lawsuits had a major subsequent impact on government and corporate actions, but owing to the strict patient certification criteria under the Pollution-Related Health Damage Compensation Law, many victims did not find redress. This led to the filing of lawsuits by 2,200 plaintiffs in seven district courts around the country over the disease definition and the responsibility of government administration. Except for the lawsuit in the Kansai area, all were canceled when the national government in May 1996 gave in and accepted a settlement proposal. Officially certified patients currently number 2,261 in Minamata and 690 in Niigata, with as many as 11,000 settling with the government. This was indeed the largest ever incident of pollution-caused health damage.

(HARADA Masazumi)

Chapter 4
Conservation and Use of Biodiversity

Rice paddies in Malaysia. Heavy use of pesticides and chemical fertilizers in recent years is impairing the role that rice paddies have played in conserving biodiversity. For this reason, traditional farming methods and methods that take advantage of natural conditions are attracting interest and being encouraged.
Photo: Ramsar Center

1. Introduction: Biodiversity as an Environmental Issue

It was only recently that people began to see the global loss of biodiversity as a serious environmental problem. In particular it was at the signing of the Convention on Biological Diversity at the 1992 UN Conference on Environment and Development (UNCED) in Rio de Janeiro where the conservation of biodiversity became an international issue. Prior to that, the matter of biodiversity was little more than a concern among certain small groups of experts like biologists, environmentalists, and environmental NGOs, who had from the 1970s been sounding warnings on the impending crisis of biodiversity loss, though the issue rarely got attention from the general public. But starting in 1992 conservation of biodiversity became a foreign policy issue, and many governments made it one of their top priority issues in their political agendas or in dealings with other countries. This change in the issue's status came about from 1990 to 1992 during the two years of negotiations for the Biodiversity Convention. In the course of those negotiations it became clear that biodiversity is a natural resource with exceedingly great actual and potential economic value. This fact attracted interest especially when at UNCED the U.S. government openly refused to sign the Biodiversity Convention because it failed to protect the economic interests of U.S. transnationals working in biotechnology and agribusiness. The refusal shocked many developing country governments that did not necessarily have a full awareness of the significance of biodiversity loss or the need to conserve it. The U.S. subsequently signed under political pressure from its own environmental organizations, but it has put off ratification because of dissatisfaction with a number of the convention's terms.

Since 1992 the public has taken a growing interest in the loss and conservation of global biodiversity, and many associated problems now receive wide news coverage. Nevertheless, the understanding of biodiversity by ordinary citizens and governments is almost always superficial, and this applies also to alarmist reports in the mass media, which have been sounding warnings on grave ecosystem damage resulting from projected species extinctions. Many governments, moreover, have only minimal understanding about the actual importance of the biodiversity issue, and that is why governments continue to pursue the kind of ecosystem-destructive plans that have heretofore depleted forests and fisheries.

Below we shall discuss the circumstances and challenges relating to biodiversity conservation in Asia, then briefly describe, and explain the significance of, eco-farming in China and wet rice aquaculture ("paddy-cum-fish farming") in Malaysia, two methods that are closely related to biodiversity conservation.

2. Conservation of Biodiversity in Asia

2-1 Biodiversity: Knowledge and Present Circumstances

One reason that many policymakers do not necessarily perceive biodiversity loss as an urgent matter is that our knowledge of biodiversity is yet uncertain. In many countries there are big gaps in our knowledge about biodiversity. For example, even species numbers and extinction probabilities are not known with any degree of certainty. In many cases biodiversity loss rates have been researched, discussed, and estimated by a few scientists in narrow fields. Especially prior to the 1980s it was nearly impossible to obtain information on biodiversity because often such information was released in the journals of specialized scientific organizations. The first systematic compilation was by the World Conservation Monitoring Centre and published in 1992 as *Global Biodiversity: Status of the Earth's Living Resources*. Another important source was the 1995 *Global Biodiversity Assessment*, a joint effort of the UN Environment Programme (UNEP) and the World Conservation Union (IUCN). These publications and other initiatives benefited from the extensive efforts of scientists around the world, but the many remaining gaps in our knowledge will require a great deal more research.

2-2 Biodiversity Conservation in Asia

Despite the lack of accurate information on biodiversity, we know enough to make the following generalizations about its conservation in Asia.

(1) Asia is a vast region with highly abundant biodiversity and it comprises all the world's main ecosystem types from the tropics to the Arctic. Many Asian countries, like Indonesia, Malaysia, and others described with the term "megadiversity," are believed to have far greater diversity and variety of habitats than other countries. Asia is also the source of many cultivated plants such as rice and other economically important plants, and has a great store of genetic resources that could improve agriculture.

(2) Asia has a high population density, as well as ancient cultures and civilizations. As a result, its vast natural environments have long been damaged and modified to suit human needs. Estimates say that at least 70% of Asia's natural ecosystems have been damaged and converted, and in some countries such as Bangladesh at least 90% of the natural ecosystems have been destroyed. Such damage and modification ultimately lead to great losses of species and their genetic resources. Natural resources are developed in accordance with the requirements of expanding populations, and thus ecosystem destruction proceeds at about the same rate as population growth.

(3) Most Asian countries are not making adequate efforts to conserve biodiversity, which can also be said for nature reserves to conserve species and for protection of those

sanctuaries. Governments' development policies emphasize economic growth at the expense of the environment, which often hinders effective efforts at conserving and protecting biodiversity, in addition to other factors such as insufficient funds and experts, and the development of natural resources at unsustainable rates. This situation clearly shows that conserving biodiversity in most Asian countries will necessitate broader international efforts and cooperation.

2-3 The Importance of Conserving Biodiversity for Sustainable Development

With the rapid loss of biodiversity there have been many attempts at assessing its importance in fulfilling the needs of present and future generations.[1] The general consensus emerging from them is that biodiversity is a critical component of sustainable development because it provides many ecological services and biological resources that contribute to meeting our future needs.

• The Importance of Habitats and Ecosystems

One aspect of biodiversity is the diversity of habitats and ecosystems in which plants and animals (including human beings) live, and on which they depend. Such biodiversity is the foundation of what are to us humans natural resources, such as forests, marine products, and many other useful products. Moreover, the ecological balance among plants, soil, and water is vital to us because it recharges water sources, and prevents soil erosion and floods.

• The Importance of Species

The reasons that other species should not be forced into extinction by human activities are not scientific and moral alone, for there are compelling economic grounds as well. Plants and animals provide us with many of our necessities such as food and medicine. We now use only a few plants and animals for these things, but there is much higher potential. We do not know, however, what plants and animals future generations will need. An illustrative example is the natural rubber tree (*Hevea brazilensis*) from the Amazon forest. Only 200 years ago this tree had no economic value, and the latex it exuded was used only to make balls for children's play. But in the 20th century it became essential for automobile tires. Similarly, we have no way of knowing what species will become the underpinning of future industries. Many organizations and multinational corporations are now screening large numbers of plant and animal species for their potential utility as industrial products. The development of technologies making it possible to analyze and synthesize useful chemical compounds has enormous economic potential.

---

[1]   See, for example:
      Fowler, G. and P. Mooney, 1990. *Shattering: Food, Policies and the Loss of Genetic Diversity*. U. of Arizona Press.
      Prescott-Allen, R. and C. Prescott-Allen, 1982. *What's Wildlife Worth?* Earthscan.
      WRI/IUCN/UNEP, 1992. *Global Biodiversity Strategy*. WRI, Washington.

• The Importance of Genetic Resources

Since the beginning of agriculture, humanity has chosen certain plants for food production in accordance with differing environmental conditions, and the variability of plants ensures that they will provide even under adverse environmental conditions. But in the course of increasing production by environmental standardization and selection for monocultures, modern agriculture is moving genetic variability in the opposite direction, which will ultimately cause the loss of genetic variability in nearly all food plants. What this means is that such plants cannot cope with changes in environmental conditions or new diseases. The awareness of decreasing genetic variability in plants whose economic value is recognized, and the awareness that the habitats of related wild plants are being destroyed, encouraged the creation of seed banks that store the varieties of economically important plants. Many of these seeds are now in safekeeping at international research institutes. Ownership and access to these varieties were at the center of heated discussion and controversy in the early 1980s at the UN Food and Agriculture Organization (FAO). Parties to the controversy, called the "seed wars," were transnational corporations, governments, and NGOs. Issues regarding the intellectual property rights of plant cultivators and breeders, and the rights of farmers, have yet to be resolved, making this a crucial task for the Biodiversity Convention, and demonstrating the economic importance of genetic resources.

2-4 The Biodiversity Convention and Asia's Circumstances

This convention resulted from the efforts of the scientific community, NGOs, UN organizations, and governments in their efforts to conserve biodiversity globally. Signing and ratifying the convention legally binds signatories to its provisions. If all the convention's goals were to be met, it would be a major contribution to the conservation of global biodiversity, and as such stakeholders are watching to see how it is implemented. Despite commendable objectives, however, implementation is fraught with problems because in many ways it inescapably manifests the latent conflicts of interest regarding priorities among governments and among domestic policies in any one country.

There is no doubt that implementation will have a broad impact on Asia because this region has most of the world's biodiversity and over half its population. Yet, Asia for the most part did not play much of a role in the convention's initial drafting process, which was mostly under the leadership of European countries and the U.S. And while the U.S. has not yet ratified the convention owing to domestic political problems, it is still very influential indirectly through its own transnational corporations, through international organizations such as the World Bank, the World Trade Organization, UN organizations, or through international agricultural research organizations under the Consultative Group on International Agricultural Research.

But cooperative relationships among Asian countries

work according to group divisions generally based on economic position, i.e., whether a certain country is a developed or developing nation. Japan, Taiwan, South Korea, and Singapore are linked with the OECD countries, while the others believe their interests are served by the developing countries. At the same time, Asian developing countries are unable to agree among themselves on a number of issues such as the conditions for use of biological resources, or funding mechanisms for biodiversity conservation.

2-5 Role of NGOs in Asia for Conserving Biodiversity

NGOs play an important role in conserving biodiversity and in the Biodiversity Convention. For example, in 1980 NGOs were the means of presenting this issue to international society, and they provided the initial information upon which convention negotiations were based. Especially well known among NGOs that performed this role were the World Conservation Union (IUCN), World Wide Fund for Nature (WWF), and the World Resources Institute (WRI). Though they originated in the developed countries, these large international environmental NGOs are sympathetic to the viewpoints of developing countries, but being environmental NGOs, they had often tended to regard the conservation of biodiversity as an end in itself. It is therefore necessary that developing country NGOs — especially those in Asia — have strategies and visions to ensure that conserving biodiversity will lead to their countries' economic development, and to assure a fair distribution of benefits to regional societies and native societies. Of course Asian NGOs have played a useful role in the Biodiversity Convention by vigorously advocating their own stands at international meetings and discussions, but with respect to organization, funding, capacity, and specialized skills, they still lack cohesion and strength, and are isolated from one another. One reason for this is Asia's historical, cultural, political, and linguistic diversity. Environmental and developmental NGOs in Asia are nearly without exception local or national, or at most regional, as with the Third World Network.

Currently there are no Asia-based international NGOs in the same sense as the IUCN, WWF, WRI, or Greenpeace, and it does not appear that any will develop to that point in the near future. Nevertheless Asia does have many NGOs, and there are many active NGOs in countries including India, Pakistan, Bangladesh, Indonesia, the Philippines, Thailand, and Malaysia. These NGOs should form international networks and pool their limited resources, which would allow them to make an effective contribution to the more appropriate implementation of the Biodiversity Convention. Following are some jobs that Asian NGOs could perform: (1) Monitor the activities of transnational corporations, especially large Japanese and South Korean corporations that have been using Asia's living resources. (2) Influence priority policy decisions. This would be possible by making policy committees take into account the concern of citizens for items such as the adverse socioeconomic effects of policies. (3) Influence and mobilize citizen opinion so that governments use appropriate methods to accommodate efforts for biodiversity conservation.

Still, NGOs have very limited opportunity to participate in COP discussions, and considering that NGOs are in fact often made light of, it is not clear how effective their activities can be. COPs are inherently venues for intergovernmental negotiations; NGOs can obtain status as observers, and that only when allowed to attend. A number of Western NGOs have through their own governments achieved progress by using this status to obtain and provide information concerning prepared resolutions, and to present critical analyses. But this approach in the West of effecting progress through governments is not necessarily effective in Asian countries where there is a long history of disputes, mutual mistrust, and antagonism between NGOs and governments. Thus if Asian NGOs are to make a contribution to conserving biodiversity, it remains to be seen how they can bring diversity to bear in their approaches.

3. Eco-Farming in China

China is dealing with crucial challenges in securing the natural resources (which include its own biodiversity) to support its economy, society, and development over the long term, but its natural resources are under heavy pressure from its 1.2 billion people and economic development based on the mass consumption and disposal of those natural resources. For those reasons China has over the last two decades made significant effort to curb environmental damage and the waste of resources, and to turn the process around.[2] The encouragement of eco-farming operations described below is part of China's effort to solve environmental problems, and is useful in preventing the qualitative decline, and providing for improvement, of the agricultural environment. While such eco-farming operations are based on ecological principles and China's traditional farming practices, they also conform to the fundamental objectives of the Biodiversity Convention, which aims to conserve biodiversity and to sustainably use its components. In fact, these operations are encouraged and implemented throughout China, and have realized promising achievements.

3-1 Current State of China's Eco-Farming

Eco-farming arose spontaneously among many farmers throughout China in the latter half of the 1970s. In the early 1980s scientists researched eco-farming and asked the government to support it. Since 1984 the central government has issued a number of documents for organizing governmental institutions to encourage and manage eco-farming. Currently there are 50 counties throughout China with

---

[2] Recently the People's Republic of China is using 0.7% of its GNP for environmental protection. While this figure is quite high for a developing country, it is insufficient for the purpose.

experimental eco-farming operations, 1,500 villages and towns, and over 20,000 eco-farming households. Except for the Tibet Autonomous Region, they can be found in all provinces, cities, towns, villages, and autonomous regions. These counties are participating members in a national eco-farming operation plan called the "Project for Constructing Fifty Eco-Farming Counties in China," which was launched in 1993 by seven divisions of the central government.[3] Within the central government there is an Eco-Farming Leading Group, of which all seven divisions are members, and in which the Ministry of Agriculture plays a leading role. Eco-farming operations receive funding from both the central and local governments, but they also generate great economic and ecological benefits for counties. Statistics show that grain production in eco-farming municipalities is 15% higher than before such operations started. Grain harvests per *mu* (0.066 hectare) have increased by 10%. Per capita income in eco-farming counties is 12% higher than other counties in the same regions, and eco-farming operations are said to have brought great environmental and ecological improvements to rural areas. Land areas with forest coverage have increased substantially, soil erosion has been alleviated, farmland soil quality has been improved, and agricultural systems are more capable of withstanding natural disasters. Eco-farming has also constituted a boost to local economies.[4] Furthermore, four of China's eco-farming villages were chosen to receive the UNEP "Global 500 Best" award for environmental protection.

3-2 Chinese Law and Policy on Eco-Farming

Under a 1992 State Council Decision, the objectives of Chinese policy for agricultural development are high production, high quality, and high efficiency, and this so-called "triple H" agriculture requires the capability of strong, sustainable support for farming ecosystems. Because eco-farming is built on a stable and balanced environment, it is a vital method for meeting the objectives of triple H agriculture.

Chinese law and policy support the development of eco-farming. Its constitution and Environmental Protection Law constitute a foundation that is favorable to eco-farming's development. For instance, Article 9 of the constitution provides that "The state ensures the rational use of natural resources and protects rare animals and plants." Article 20 of China's Environmental Protection Law states, "[T]he people's governments at various levels shall provide better protection for the agricultural environment by preventing and controlling soil pollution, the desertification and alkalization of land, the impoverishment of soil, the deterioration of land into marshes, earth subsidence, damage to veg-

etation, soil erosion, the drying up of sources of water, the extinction of species, and the occurrence and development of other ecological imbalances, by extending the scale of comprehensive prevention and control of plant diseases and insect pests, and by promoting a rational application of chemical fertilizers, pesticides, and plant growth hormones."

China's Agriculture Law has a chapter on the protection of agricultural resources and environments, and it sets forth a number of requirements directly relating to eco-farming. Some of these are agricultural resource plans, local energy development plans, plans for ecosystem/environment use, protection and conservation of farms, fertilizer use, restricting pollution, controlling soil erosion, and reforestation. This law is the basis for providing funds to projects that protect agricultural environments, and it supports eco-farming. In June 1985 the State Council released a document entitled "Suggestions Concerning the Development of Eco-Farming and the Strengthened Protection of the Agricultural Eco-Environment," which called on local organizations to develop eco-farming in accordance with the ecological conditions of their localities. In August 1996 the State Council released a decision on environmental protection that instructed local governments to develop eco-farming, and to prevent and control pollution of agricultural environments.

Eco-farming also enjoys the support of many local laws throughout China. For example, Shanxi Province's 1991 Regulations on Agricultural Environmental Protection set forth "the development of eco-farming and the sensible use of agricultural resources" as one responsibility of local governments. Hebei Province's 1993 Agricultural Environment Protection Regulations have similar provisions.

3-3 Types of Eco-Farming Operations

There are many types of eco-farming in China because of the country's vast land area and broad variety of natural conditions. Each eco-farming operation must accord with its local natural conditions, and economic development priorities differ from one locale to another. Thus the form taken by eco-farming ordinarily reflects local priorities for economic development.

Eco-farming can be categorized into a number of groups according to different indicators. China's National Environmental Protection Agency divides eco-farming into the following five categories.[5] (1) Eco-farming based on symbiotic organisms, (2) eco-farming operations based on the recycling and reuse of materials, (3) eco-farming operations based on the use of natural enemies to control insect pests, (4) eco-farming meant to deal with important problems (eco-farming operations for preventing desertification and improving soil), and (5) eco-farming operations integrated into com-

[3] Ministry of Agriculture, State Planning Commission, State Science and Technology Commission, Ministry of Finance, Ministry of Forestry, Ministry of Water Conservancy, and the National Environmental Protection Agency.

[4] The foregoing facts and figures come from *Report on Environmental Protection in the People's Republic of China*, released in June 1996 by the State Council's Department of Information.

[5] For details see: National Environmental Protection Agency and United Nations Environment Programme, *China's Eco-Farming*, China Environmental Science Press, Beijing, 1992.

prehensive regional development plans. These are further broken down into subcategories (Table 1).

3-4 Case Studies of Eco-Farming Operations

As mentioned above, four operations won UNEP Global 500 Best awards.[6] Let us examine two of them briefly.

• Liuminying Village (Daxing County, Beijing)

Liuminying is a typical farming village on northern China's Huabei Plain. Its eco-farming operation consists in the recycling and reuse of materials by integrating grain cultivation, aquaculture, animal husbandry, biogas use, and processing. Villagers use grain plant stalks and the bran from rice and wheat as animal fodder. Manure and crop stalks are used as feedstock for biogas digesters, which supply farmhouses with energy. Dregs and water from the biogas digesters are processed into fodder for fish and pigs, and also

_____

[6] Liuminying Village in Daxing County (Beijing, 1987), Shanyi Village in Xiaoshan City (Zhejiang Province, 1988), Heheng Village in Taixian County (Jiangsu Province, 1990), and Xiaozhang Village in Yingshang County (Anhui Province, 1991).

used as fertilizer on fields and orchards. Mash left over from making soy milk is used as fodder for milk cows and pigs. Chicken manure is used as pig feed or fertilizer. This total reuse and recycling of materials increased the production of grains, beef, milk, pork, chicken, eggs, ducks, fish, and vegetables, decreased coal consumption and pollution, and improved the quality and environments of farms. The village's economic conditions improved, and farmers' incomes increased substantially.

• Xiaozhang Village (Anhui Province, Yingshang County)

Located in the southern part of the Huabei Plain, this village consists of 690 households and 2,970 people on 422 hectares. Its eco-farming operation consists in the integrated development of agriculture, forestry, livestock, aquaculture, and associated industries. The village began its eco-farming operation by taking advantage of its low-lying topography and water. Villagers built 20 ditches and water channels, reclaimed 16 hectares, built 15 fish ponds, dug pumped wells in 64 locations, and built two irrigation and drainage pumping stations. Interplanting and mixed cropping increased the solar energy use of farms and the production of grains and economic crops. Villagers planted tree shelterbelts

Table 1 Eco-Farming Categories in China

A. Eco-farming based on symbiosis
  1. Three-dimensional cultivation
  2. Mixed animal husbandry
  3. Combining livestock and crop cultivation
B. Eco-farming operations based on the recycling and reuse of materials
  1. Recycling and reuse of materials in growing grains
  2. Recycling and reuse of materials in livestock raising
  3. Recycling and reuse of materials by integrating grain cultivation and livestock raising
  4. Recycling and reuse of materials by integrating grain cultivation with livestock raising and processing industries
  5. Recycling and reuse of materials by integrating grain cultivation with livestock raising and biogas use
  6. Recycling and reuse of materials by integrating grain cultivation, aquaculture, livestock raising, biogas use, and processing industries
C. Eco-farming operations based on the use of natural enemies to control insects
  1. Using natural insect enemies to control insect pests
  2. Using birds to control insect pests
  3. Using grass and herbs to control weeds and insect pests
  4. Using fungi to control insect pests
D. Eco-farming meant to deal with important problems (eco-farming operations for preventing desertification and improving soil)
  1. Preventing desertification
  2. Preventing soil erosion
  3. Soil improvement
E. Eco-farming operations integrated into comprehensive regional development plans
  1. Combination of agriculture, forestry, and livestock raising, with forestry being the main element
  2. Encouraging basic industries by resolving energy shortages
  3. Comprehensive management of water, soil, forestry, and equipment resources mainly in accordance with livestock raising
  4. Facilitation of agriculture, forestry, industry, commerce, and transportation through livestock and poultry raising
  5. Integrated development of agriculture, forestry, livestock, aquaculture, and related industries
  6. Comprehensive planning of eco-farming operations

with a total length of 40.8 km. Trees were also planted on the village periphery and on the banks of ditches and along roads. In afforestation villagers decided to mix tree species, so plantings include combinations of trees, shrubs, and grasses, deciduous and evergreen trees, and broadleaf trees with conifers. They also planted 28.7 hectares of orchards and 8 hectares of bamboo groves. In all, the village planted 167,000 trees, which comes to 56 for each villager. Forested area increased from 7.6% to 26.9% of total village area, for timber storage of 4,070 cubic meters. In one year, the 175,000 kilograms of pruned branches serve as firewood, and over 1.3 million kg of fallen leaves are gathered and used as fodder and in making compost. Villagers keep pigs, cattle, sheep, rabbits, and other animals in the enlarged pastures, and raise freshwater fish in fish ponds. These livestock and aquaculture operations provide manure fertilizer for the farms.

Livestock manure is used to generate biogas. This village considers bioenergy important and has developed highly efficient gas stoves. It built 48 biogas digesters that supply energy to all village farmhouses. The use of biogas digesters and the energy-efficient stoves economizes 1.4 million kilograms of grain stalks and grass yearly, which can be returned to the fields and used as livestock feed. Xiaozhang Village is also developing associated industries such as those to process farm produce. Eco-farming operations like these have improved natural environments and raised living standards, so that per capita annual income rose from under 100 yuan in the 1970s to 1,424 yuan in 1989.

As these examples illustrate, eco-farming operations in China have made progress and considerable achievements over the last two decades, which will serve as a solid foundation for future advances. These operations benefit not only the economy, but also environmental conservation, conservation of biodiversity, and the sustainable use of living resources. Nevertheless, further progress requires the elimination of several difficulties and problems that constitute impediments. For example, China's eco-farming operations heretofore have nearly all been on the household or village level. How to develop eco-farming operations on a larger scale is a matter of regional land use planning and environmental protection, so China should henceforth find ways to reconcile eco-farming operations with these concerns. Furthermore, governments on all levels should invest more human and fiscal resources in studies and experiments for eco-farming. China must further strengthen government agencies concerned with the management and protection of eco-farming, and adopt policies that further encourage eco-farming. It should prepare more funds for eco-farming and provide more technical support. China should also take steps to support farmers in the sale of eco-farming produce. It will be necessary to train more farmers in order to increase the practice of eco-farming. Another important job will be to explore the connections between eco-farming and biodiversity, especially the relationship with the Biodiversity Convention's provisions and guidelines.

4. Wet Rice Aquaculture in Malaysian Rice Paddies

This section is a brief overview of Malaysian wet rice aquaculture and its significance.

4-1 Malaysia's Paddy-cum-Fish Farming

To keep pace with population increases, people developed the rice paddy system of irrigation to replace the traditional method of depending on rainfall. Malaysia's main rice-producing regions have built holding ponds to supply water to paddies, and also irrigate by pumping water from rivers. Principal rice-producing regions therefore present a new kind of habitat consisting of water channel networks around the paddy fields for irrigation and drainage. These channels give fish a place to live during times when paddies are untilled and dry. When the rainy season begins and water levels in the paddies rise, adult fish start laying eggs in preparation for the next generation. Ecosystems of areas irrigated in this manner consist in the close relationships among paddies and their irrigation and drainage channels, and the rivers and holding ponds to which they are connected. Even in traditional paddy ecosystems, fish have over many years adapted to environmental conditions with the two extremes in which paddy fields are either dry or flooded, but fish living in these ecosystems have made even bolder and more clever adaptations in order to survive.

4-2 Paddy Characteristics, Flora, and Fauna

Traditional paddy ecosystems exposed life there to extreme environmental conditions such as high-temperature, little dissolved oxygen, high humidity, and low water level, but modern rice-growing techniques are now highly mechanized, thus adding various anthropogenic pressures. For instance, although farmers are quite judicious in their use of chemical fertilizers, pesticides, and herbicides, these are used even now in accordance with necessity because producing rice is the objective. In Malaysia's principal rice-producing districts, most rice ecosystems are a rice monoculture. The environmental conditions of these ecosystems can be summarized as follows: (1) Shallow water of 30-50 centimeters in depth, (2) high temperatures of 26-40°C and large temperature difference, (3) large 0-10% fluctuation in dissolved oxygen concentration, (4) dryout of soil between rice harvests, (5) nutrient supply from fertilizers, (6) 10-60% alkaline content, 0.3-0.8% nitrates, 0.02-0.8% orthophosphoric acid, and (7) application of various herbicides, pesticides, and other chemicals as rice plants are growing.

These environmental conditions make for unique flora and fauna in Malaysia's rice paddies. There are aquatic plants such as phytoplankton, algae, submerged plants, floating plants, and emerging plants. Submerged plant species and suspended green algae serve as food for fish, and also provide good places for them to lay eggs and raise their young. Floating plants are regarded as weeds, and many are found in paddies. By proliferating they promote excessive evaporation of paddy water, and when growing thickly they prevent fertilizer from dissolving in the water. But as with one kind of sweet potato, they are disliked when flourishing but some are edible. Some of the emerging plants are also edible. Paddies are also home to many kinds of zooplankton, which are an important food source for fry, while juvenile fish feed on the many aquatic insects found in paddies. Other insect inhabitants include stink bugs, dragonflies, damselflies, mosquitoes, flies, and horseflies, which not only feed the fish, but also help control insect pests. Fish are the most important inhabitants of Malaysia's rice paddies, and ordinarily at least 10 species are found. All of them are robust by nature and can endure the large physical and chemical fluctuations of paddy environments. Fish are important because they not only serve as food for the farmers, but also control insect pests. Many wild birds also visit rice paddies. Many migratory birds use the paddies as staging areas. The highly diverse environments of paddies are important for birds because it is easy to find food there. These rice-growing ecosystems, which include nipa palm marshes, riparian vegetation along water channels and rivers, wetland vegetation, and the like, are used by birds as foraging areas and refuge.

4-3 Development of Wet Rice Aquaculture

Malaysia's rice aquaculture came from India and into Southeast Asia about 2,000 years ago. Aquaculture in rice paddies started when people raised the wild fish in paddies by putting them inside crawls. In the beginning the farmers simply caught fish in traps and kept them until the paddies dried out at the rice harvest. At first fish were consumed locally, but later farmers began selling fish for cash income, which led them to make a number of improvements in the capture and raising of fish to increase production, and this became the rice aquaculture that is still practiced in basically the same way now. People have raised efficiency by building ponds, depressions, and other places where fish can take refuge when water levels fall and paddies become inhospitable. These places further enlarged fish habitat. Some people enlarged fish refuges by digging deep channels around paddy fields. Fish that wander into the channels during floods are caught there and raised in the paddies until the rice crop finishes and the fish are harvested. Ordinarily the fish require no feeding in particular because they depend for nearly all their food on things that are naturally found in the paddies. In some instances, however, people feed the fish kitchen scraps, leftovers, chicken entrails, and other things the fish will eat. To use land effectively, some farmers worked out a cleverer farming system, which involves planting squash, beans, okra, eggplant, sweet potatoes, corn, yams, sugar cane, and other crops on the banks of the ponds and channels where fish take seasonal refuge, and using them as places to raise ducks and other domestic fowl because bird droppings raise paddy output. The fish are a precious source of cash income to farmers, who in one season make 66 to 100 ringgit. This is all profit because the farmers invest hardly anything in raising fish.

4-4 The Future of Paddy-cum-Fish Farming

In highly mechanized rice-growing areas of Malaysia it is now difficult to manage this system of farming, but there are considerable hopes for its future potential in view of its positive significance as a sustainable agricultural system. For example, as the farmers themselves eat the fish that grow in their paddies, they started to limit their own use of pesticides and herbicides, and for the sake of the fish, which are a valuable resource, the farmers now use these chemicals only when especially needed, such as to control unusual outbreaks of insect pests and algae. Furthermore, our research demonstrated that the decreased use of pesticides and herbicides has had no effect on rice yields. In fact the very presence of fish helps control large outbreaks of insect pests and algae. Moreover, reduced use of insecticides and herbicides has the positive benefit of facilitating the growth of paddy fish, resulting in greater catches of fish from paddies and water channels. In the future there may be plans for aquaculture operations to attain greater fish density in the channels, and in doing so the use of native fish species will be better than introducing exotic species because native species are more robust and demand a higher market price. This would be effective in achieving progress in the conservation and management of biological diversity throughout Malaysia.

Finally, rice paddies and their water channels are now popular as places for people to enjoy fishing, and on weekends one can find fishing enthusiasts gathered there. This helps people understand the benefit of rice aquaculture, and will lead to better management and wiser use of rice paddies. Raising public awareness toward a positive evaluation of this farming system will no doubt guarantee the future conservation and sustainable use of rice paddies, which are extremely valuable wetland ecosystems.

5. Summation

Interest is focused on the protection of species and the use of genetic resources in connection with biodiversity, but one of the purposes of the Convention on Biological Diversity is the sustainable use of biodiversity's components, and in particular emphasizes the application of traditional forms of use.

As we saw in these examples from China and Malaysia, in Asia these elements are widely used in primary industries such as agriculture, forestry, and fisheries. But the rapid modernization of Asian countries brings changes in the traditional forms of utilization and management, which is impairing biodiversity.

Thus conserving biodiversity requires that we turn our attention not only to wilderness, but also to human-altered nature. Such nature conservation and management depends on the actions of people in local primary industries. Nevertheless, the difficulty of completely returning to traditional forms of use requires that we establish new human-nature relationships that do not diminish biodiversity, while supporting such efforts will necessitate economic and legal measures on the national and international levels.

(LEONG Yueh Kwong, WANG Xi, AHYAUDIN bin Ali, ISOZAKI Hiroji, MARUTA Sayaka, NAKAMURA Reiko)

Essay 1　Per Capita Natural Resources in China

In terms of the absolute amounts of natural resources, the People's Republic of China is one of the world's richest countries, but dividing those abundant resources by its population of 1.2 billion makes China one of the world's poorest nations because it has very few resources per person. For example, the area of arable land per Chinese is only 0.086 hectare, which is one-fourth the world average of arable land per person, 0.344 hectare. The bare minimum per capita arable land area needed is said to be 0.053 hectare. Each Chinese has 0.133 hectare of forested land, which is 11.3% of the world average. China is seventh-place in the world with regard to total water flow, but that amount is only one-fourth the world average in per capita terms. China's per capita mineral reserves are half the world average, and 80th place worldwide.

(WANG Xi)

Essay 2　Asia's Wetlands and Rice Paddies

Wetlands are the most important aquatic ecosystems, and among them rice paddies are a special habitat because they are seasonally flooded and dry. Primitive rice paddies were originally flooded with water when the rains came, and in that sense are an extension of marsh ecosystems. For that reason they host the same kinds of flora and fauna as marshlands. Biodiversity in paddies is principally defined by the existence of species that are uniquely equipped to cope with these special habitats, which are alternately dry and flooded. Many mobile aquatic organisms like fish enter and leave paddies in accordance with the rainy and dry seasons, and that is why we regard paddy fields as marshlands. Even now we can find primitive paddies — marshland ecosystems — that depend on rainfall to flood them. For example, paddies in Cambodia and around the Mekong River become marshes when the monsoons come. At this time the farmers catch fish that have strayed into the paddies and eat them or sell them for supplementary income.

No matter how they are flooded, Asian paddy fields play an important role with wetlands in conserving marsh ecosystems.

(AHYAUDIN bin Ali)

Part II   Asia by Country and Region

Chapter 1
Japan

1. Introduction

Written records of pollution and environmental problems in Japan can be found from as far back as the 1600s, but it was not until the latter half of the 1800s that such problems arose routinely in connection with economic activities, and that especially in the 1880s when the Industrial Revolution came to Japan. Already in the 1870s the Japanese word for "pollution" had been coined, and by the first two decades of the 1900s it was being used as a catch-all term for air, water, and noise pollution, vibration, unpleasant odors, and other things detrimental to public health. Some of the main instances of prewar pollution were water contamination by the Ashio copper mine, sulfurous acid gas from copper smelting in Besshi, Hitachi, and Kosaka, and air and water pollution in places including Osaka, Amagasaki, and Kawasaki. These and other frequently occurring problems made political waves at that time. Among mainly victimized farmers there was a great deal of opposition and they carried out campaigns for redress, in some cases lasting as long as 50 years. Thus already in the prewar years, anti-pollution movements had devised all the currently conceivable principles for dealing with pollution, such as stopping sources, dispersion and dilution using high smokestacks and other means, scattering factories instead of building in clusters, and emergency measures enacted as when meteorological conditions worsen.

Especially in the late 1920s and 1930s there were advances in research on pollution control technologies. One landmark achievement was the world's first practical flue gas desulfurization by Sumitomo Metal Mining in 1934, which was induced by pressure from local citizens. Compensation won by pollution victims, which would amount to about 10 billion yen when converted to present monetary value, was all used to build public facilities for the local community instead of being divvied up among the victims for personal use.

Quite unfortunately, however, the corporations and the national and local governments completely forgot about these valuable experiences during the economic growth period in the Second World War and thereafter, and neglected to take the necessary measures against pollution. As a consequence, many areas of postwar Japan became the scenes of serious pollution and environmental damage that were nearly unprecedented worldwide. Below we shall start by reviewing pollution and other environmental problems in postwar Japan.

2. Brief History of Environmental Problems in Postwar Japan

Postwar Japan's environmental problems can be divided into five time periods, each of which is examined briefly below.

2-1 Rapid Growth and Serious Pollution (1954-1964)

After World War II, especially during the two decades beginning in 1954, Japan's economy took the express lane to establishing the heavy and chemical industries, and to building big cities. Although heavy pollutant emissions were naturally expected, both the public and private sectors cut expenditures by holding down investment in pollution control and environmental protection, which resulted in pollution that had hardly been matched anywhere else in the world at any time. During 1960 Osaka had 156 smog days, and the one-day average SO_2 concentration in that city's Nishiyodogawa area was 0.189 ppm (the highest was 0.321 ppm), which was bad enough to produce air pollution victims. Around 1965 the biological oxygen demand (BOD) near Kyobashi in the Yodo River, which flows through central Osaka, reached 40 ppm, making it more like a cesspool than a river. Along the bay shoreline the pumping of groundwater and gas caused rapid ground subsidence with a maximum cumulative subsidence since 1935 of 250 cm, causing 2 square km of the Nishiyodogawa area to sink into the sea. Much of what happened in Osaka could be found in all Japan's large cities in those days.

During this period, pollution from the prewar years carried over and caused two outbreaks of Minamata disease (Kumamoto and Niigata prefectures) and itai-itai disease (Toyama Prefecture). Additionally, "Yokkaichi-type" pollution occurred at the petrochemical complexes, which were on the cutting edge of technological innovation in postwar Japan. The central government facilitated rapid economic growth by siting these complexes on the coast adjoining major urban areas, and implemented nodal development in outlying cities on the basis of the First Comprehensive National Development Plan, which led to the spread of Yokkaichi-type pollution all over the country.

Meanwhile, pollution control ordinances were passed in Tokyo, Osaka Prefecture, and Fukuoka Prefecture beginning in the 1950s, but they lacked teeth because standards were lax and corporations resisted. In 1958 the government enacted two water pollution laws, which were the nation's first pollution-control laws, but these did not apply to Minamata Bay, where the victims of mercury contamination had begun appearing in 1956. In fact, the government stifled the organic mercury theory proposed by a Kumamoto University research group in 1959. It was not until 1968, when the production of acetaldehyde had halted, that the government applied the two water pollution laws and admitted to the health damage, but that was too late to stop the second occurrence of Minamata disease in Niigata Prefecture in 1964.

In Yokkaichi Port near the Yokkaichi petrochemical complex around 1958, people were catching oily-smelling fish that had been contaminated by water pollution, and air pollution victims began appearing in the area about 1959, but little was done to deal with the situation. The government enacted the Smoke Control Law in 1962, but its weak standards allowed emission concentrations higher than those for the flue gas desulfurization already achieved by Sumitomo Metal Mining before World War II, which allowed pollution-caused harm to become even more widespread.

2-2 Public Opinion and Citizens' Movements Oppose Pollution — Progressive Local Governments and Pollution Lawsuits (1964-1970)

In 1970 the ratio of the heavy and chemical industries in Japan's total manufacturing industries attained 62.4%, which far exceeded the 56.5% of second-ranked West Germany. This made for very large pollution loads of 50.2 kg SO_2 and 29.3 kg BOD per 1 million yen in manufactured goods shipped. These figures were much larger than the 40.6 kg SO_2 and 21.0 kg BOD of West Germany at that time. Private-sector investment in pollution control in 1970 was 188.3 billion yen, or 5.3% of total investment in plant and equipment. While this seems like a high percentage, Japan's private sector had in those days at last begun pollution-control measures owing to the 1967 Basic Law for Environmental Pollution Control. If one takes water pollution control as an example, in 1970 a mere 8.9% of BOD was being removed.

From 1960 to 1970 the population of the three major urban areas of Tokyo, Osaka, and Nagoya increased from 37.4 million to 48.3 million, while nearly all the oil refining, petrochemical, and steel production facilities, which were the mainstay of the heavy and chemical industries, were concentrated in these same three areas. SOx emissions per square km of habitable land area increased from 16.2 tons in 1955 to 131.3 tons in 1971, with the latter amount being three times the national 1971 average of 45.6 tons. Indeed the central portion of Japan was unfit for human habitation.

In 1963 and 1964, as pollution was intensifying around these complexes, a citizens' opposition movement arose in the Mishima-Numazu-Shimizu area of Shizuoka Prefecture, where the government was planning a petrochemical complex. Calling for "no more Yokkaichi," the opposition engaged local scientists to perform environmental impact assessments and other studies. Citizens also held over 300 meetings to study the problem. A new feature of this campaign was that, instead of complaining to the central government as was usual, citizens focused their efforts locally by reforming the local government. Taking this seriously, the national government conducted the first environmental impact assessment in Japan by a government survey group and tried to head off the local opposition movement, but in the end the government's attempt failed. This Mishima-Numazu-Shimizu campaign was the first full-fledged citizens' movement in Japan, and its success roused anti-pollution public opinion and triggered movements around the nation. The public opinion and movements later paved the way — in two directions — for the advance of Japan's own unique pollution and environmental policy.

First, in areas where most people were pro-environment, the public elected environmentally minded, reformist politicians to local offices. These politicians implemented strong administrative guidance based on pollution-control agreements and ordinances, and other tough anti-pollution measures, and also sought a shift in the central government's pollution and environmental policy. Especially well known is Tokyo's 1969 pollution control ordinance. As this ordinance was tougher than national laws, the central government claimed it was illegal and disputed it, but ultimately knuckled under in the face of both international and domestic anti-pollution public opinion, not to mention the severity of pollution at that time. This state of affairs forced the government in late 1970 to hold the so-called "pollution Diet session," which revised the 1967 Basic Law for Environmental Pollution Control and passed 14 other pollution-related laws.

The other policy direction involved the filing of pollution lawsuits in areas of the country where environmental sentiment was relatively weak, and pollution victims found themselves isolated. In close succession victims filed the "Four Great Pollution Lawsuits": Minamata disease in Kumamoto and Niigata prefectures, itai-itai disease, and Yokkaichi pollution, all of which ended in total plaintiff victories in the early 1970s. Victory in the Yokkaichi case in July 1972 had an especially great influence because it led to the total revision of environmental quality standards for air and other types of pollution, and to passage of the 1973 Pollution-Related Health Damage Compensation Law (below, Health Compensation Law).

2-3 Progress in Coping with Pollution (1970-1978)

During the early 1970s, therefore, there was a measure of progress in resolving some kinds of industrial pollution. The world's toughest environmental quality standards (EQS) were established on substances such as SOx, organic mercury, cadmium, and lead that had caused especially serious pollution damage, in response to which the private sector in 1975 invested 964.5 billion yen (17.7% of total capital outlays) in pollution control. The OECD's 1977 review of Japan's environmental policy stated that although Japan had won victories in many battles to eliminate pollution, it had yet to win the war to improve environmental quality. Indeed, some of the visible pollution-caused harm had by that time disappeared from view.

But from that time the nature of Japan's environmental problems began to undergo major change because society faced new problems in forms that included environmental and pollution damage caused by public works construction for airports, roads, trains, and the like; pollution by the huge volumes of wastes produced by mass production and mass consumption; and nuclear power accidents. Additionally, the public found new concern with issues such as the conservation of forests, farmland, beaches, and the oceans, as well as environmental quality issues like conserving scenic areas and historical areas of cities. Especially in regard to the latter category, Japan's environmental policy lags far behind, as the OECD has observed.

A landmark achievement of these years that deserves special mention is the control of emissions from gasoline-powered passenger cars. The government adopted a policy under which a strict standard (cutting NOx emission concentration to one-tenth) would be completely attained in 1978. This achievement was brought about by public opinion against automobile pollution, and the efforts of a survey group that visited Tokyo and six other large urban centers. What is more, the Osaka High Court handed down a landmark decision in the lawsuit against Osaka Airport noise pollution, recognizing all three of the plaintiffs' demands: payment of damages, injunction on night flights, and payment of future damages until remedial measures are taken against pollution. Under government pressure the Supreme Court overturned this decision in 1981, but it finalized the payment of damages. This overthrew the nationalistic thinking that had held sway since prewar days that public works were for the public weal, and that people should therefore put up with any pollution and human rights violations that might result. Here was the beginning of an era when the public sought to establish a principle under which pollution and environmental damage would not be permitted even for public works.

2-4 Regression of Environmental Policy (1978-1988)

Under the world structural recession that continued after the 1973 oil shock, the regression of environmental policy began in many countries. In Japan this did not happen right away, as seen from the stricter controls on motor vehicle emissions, but there was gradually stronger pressure on environmental policy from industry and conservative politicians. Especially at the 1977 London Summit there was criticism of Japan's export-based economic policy, and demands that Japan expand domestic demand, to which Japan responded by beginning a policy swing back toward loosening environmental controls. For example, decisions were made to resume construction of the Seto Bridge and expressways, which had halted due to anti-pollution movements. And since the world's strictest NO_2 EQS of 0.02 ppm for a daily average had hampered these projects, the Environment Agency relaxed it substantially to 0.04-0.06 ppm. Only one week later, the decision was made to begin work on the Seto Bridge. Then in the late 1970s the major reform-minded governors in Tokyo, Osaka, Kyoto, Okinawa, and other places left office, starting a swing back toward conservatism in regional politics. Although there were exceptions, citizens' environmental campaigns gradually came to a standstill, and after the Supreme Court decision in the Osaka Airport lawsuit mentioned above, the judiciary likewise began to retrogress from its role, as in the first half of the 1970s, of overseeing the legislative and administrative areas with priority for the environment; there was a gradual tendency to grant priority to development and to rubber-stamp the intentions of government administrators. The changes leading to these new circumstances were not unrelated to the appearance in the West, beginning in the latter half of the 1970s, of the neoliberalism (i.e., new conservatism) that advocated privatization, deregulation, and "small government." With the birth of the Nakasone Cabinet in the early 1980s, that political trend reigned in Japan, too. One manifestation of that trend was the total revision in 1988 of the Pollution-Related Health Damage Compensation Program (Health Compensation Program; see **Essay 2**), which, beginning in March of that year, canceled the regional designations for air pollution and halted the certification of new air pollution victims. During these years the NO_2 concentration in Japan's major urban areas actually increased, and air pollution tended to steadily worsen.

2-5 Internationalization of Environmental Issues and Policy (1989-)

Around the time of the 1992 Rio Earth Summit, "foreign pressure" induced modest progress in Japan's environmental policy. It was a period when Japan's industrial structure shifted its emphasis from the heavy and chemical industries to hi-tech, information, and services. A variety of changes also occurred in relation to pollution and the environment. The transformation of Japanese companies into transnational corporations, and increasing official development assistance (ODA) by the government raised pointed questions about Japan's international responsibility for environmental problems in developing countries, particularly Asian ones. For these reasons it is important now that citizens' environmental movements in Japan likewise internationalize and build solidarity with those in other countries, especially in Asia. Society has taken up the new challenge of sustainable development that was proposed at the Earth Summit. In view of Japan's history and its experiences to date, there are strong expectations that in order to genuinely respond to this new challenge the Japanese will build grassroots environmental public opinion and campaigns, and that local authorities will once again take action in accordance with their independent and original thinking.

3. Development of Environmental Law and Policy

This section will briefly review the historical development of Japan's environmental law and policy.

3-1 Early Environmental Law and Policy

This era can be divided into two periods: (1) the period from 1868 to the years before WWII, and (2) the postwar years until about the end of the 1950s. During the prewar period there were serious pollution incidents like the aforementioned water contamination by the Ashio copper mine, but in those days Japan still had no pollution laws or effective ways to deal with pollution. Factors working against environmental legislation and controls were the "enrich the country and strengthen the military" policy adopted in the 1860s, and the undeveloped state of democracy, human rights, and local autonomy. But after the war Japan managed to create environmental laws in the real sense of the word, which began with local government ordinances (pollution control ordinances were enacted by Tokyo in 1949, Osaka Prefecture in 1950, and Kanagawa Prefecture in 1951). By contrast, the national government did not embark on pollution control legislation until the end of the 1950s.

The water pollution incident that occurred in 1958 at Urayasu on the lower reaches of the Edo River, which runs between Tokyo and Chiba Prefecture, triggered the first pollution control legislation by the national government: the Water Quality Preservation Law and the Factory Effluent Control Law, collectively called the two water quality laws. And as respiratory illnesses grew more serious in the new industrial cities such as Yokkaichi, where the illness was called "Yokkaichi asthma," the government became aware of the need to pass laws controlling air pollution, and in 1962 it passed the Smoke Control Law. These laws were, however, woefully inadequate in terms of both provisions and implementation, and were totally powerless to stop the spread of health damage like Minamata disease or Yokkaichi asthma.

3-2 Creation of Environmental Law and Policy

• Pollution Control Laws

In Japan the word for "pollution" (*kougai*) has been used to signify damage to human health, flora and fauna, and property by contamination of the environment. This damage grew serious in the 1950s and even more in the 1960s, resulting in repeated grave types of health damage as symbolized by Minamata disease, itai-itai disease, and Yokkaichi asthma. The rising tide of public calls for concrete action finally made the central government see the need for concerted action instead of the fractionalized measures taken by individual ministries and agencies until then. Efforts toward a basic law began and led to passage of the 1967 Basic Law for Environmental Pollution Control. This law defined what were called the "seven typical pollution types" of air pollution, water pollution, soil contamination, noise, vibration, ground subsidence, and offensive odors (soil contamination was added in a 1970 revision), created a system of policy measures that clarified the duties of businesses, the national and local governments, and the citizens, and on this basis provided for the passage of individual pollution laws. Although Japan had no single government agency specializing in pollution, this law did establish the basis for the system of pollution laws by which agencies had to abide. The task then was to create the individual laws that would translate those policy measures into action. In 1968 came the Air Pollution Control Law to take the place of the 1962 Smoke Control Law, as well as the Noise Regulation Law to

deal with steadily increasing noise. But as these laws were not necessarily adequate to the task, pollution grew progressively worse and more diverse.

Meanwhile, the lawsuits against Niigata Minamata disease, Yokkaichi asthma, itai-itai disease, and Kumamoto Minamata disease brought immense pressure to bear on the central government. In late 1970 the "pollution Diet session" passed the Water Pollution Control Law, which took the place of the previously mentioned "two water pollution laws." The passage of new laws or revision of existing laws resulted in a set of 14 pollution laws, to which was added the Offensive Odor Control Law in 1971. In this way the government created Japan's system of environmental laws and policy consisting of the Basic Law for Environmental Pollution Control as the foundation, upon which rest the individual laws that implement pollution control.

Nevertheless, this did not mean that pollution had been whipped. As with areawide total pollutant load control for SO_2 and NO_2, certain progressive local governments proceeded with their own effective measures that did not find their way into national laws until several years later.

• Nature Conservation Laws

Another important problem was how to conserve the natural environment. Except for protection of national parks and other natural parks by the 1957 Natural Parks Law, Japan at that time had no effective legal institutions for conservation. But the 1970 revision of the Basic Law for Environmental Pollution Control established a new provision that obligated the government to protect the natural environment. The Environment Agency, established in 1971, also had an administrative department for environmental protection. Nature conservation measures therefore started progressing along with those for pollution control. Enactment of the Nature Conservation Law in 1972 set forth a clearly defined philosophy and basic policy for nature conservation. The government began working out concrete policy measures, and, despite inadequacies, to expand the system of environmental law from pollution control to include environmental protection as well, with the result being a system characterized by and grounded in the philosophies and basic policies defined by the Basic Law for Environmental Pollution Control and the Nature Conservation Law. This system remained in place until 1993, when the Basic Environment Law (see below) was established. Nevertheless, many items closely connected with environment-related administration remained in the provinces of the Ministry of International Trade and Industry (MITI), the Ministry of Construction, the Ministry of Agriculture, and other business/project agencies.[1] Even after founding the Environment Agency in 1971, therefore, Japan was basically unable to eliminate the harm arising from compartmentalized control over environmental matters.

---

[1]   Central government ministries and agencies concerned with promoting business, public works projects, and the like.

3-3 Stagnation and Regression in Environmental Law and Policy

While environmental law and policy were able to curb pollution to an extent, around 1973-74 there came the new challenge of creating laws to prevent pollution and environmental damage. Legislating an environmental impact assessment system would be an example. But despite several years of legislative efforts and environmental administrators' political compromises to help those efforts (which watered down the EIA bill), opposition from industry and business/project agencies killed it, leaving the matter to be addressed by a 1984 Cabinet decision that was also highly inadequate. In 1978 the government had relaxed the NO_2 EQS, and a decade later it totally canceled class 1 regions designated under the Health Compensation Law for air pollution. Although SO_2 pollution had indeed been mitigated, that was decidedly not the case for NO_2, which was actually worsening along roads, and especially with increasing diesel exhaust, it was clearly too early to cancel the designations that were meant to help air pollution victims. These were years of serious regression in environmental policy.

This is not to say there was no progress at this time. the 1973 Law Concerning Temporary Measures for Conservation of the Seto Inland Sea was made into the permanent Law Concerning Special Measures for Conservation of the Seto Inland Sea in 1978, and the Law Concerning Special Measures for the Preservation of Lake Water Quality was established in 1984 in response to deteriorating lake water quality. A noteworthy development at this time was that the sphere of environmental issues expanded to subsume also the preservation of "amenities,"[2] a change induced by proposals from citizens and the spread of environmental administrative efforts at the local government level. Since then amenity has gradually come to hold an important place at the national level as well.

3-4 New Progress in Environmental Law and Policy

In the 1990s environmental law in Japan started making new progress, which was catalyzed by the rising awareness of the global environmental crisis, specifically the 1992 Rio Summit. Although environmental policy had stagnated, the new decade brought revision of the Wastes Disposal and Public Cleansing Law and establishment of the Law for the Promotion of Utilization of Recycled Resources ("Recycling Law") in 1991, passage of the Law for the Conservation of Endangered Species of Wild Fauna and Flora in 1992, and more. Most significant is the Basic Environment Law established in November 1993. It set up three fundamental principles as the philosophy of environmental policy: (1) "Enjoyment and Future Success of Environmental Blessings" (Article 3), (2) "Creation of a Society Ensuring Sustainable Development with Reduced Environmental Burden" (Article 4), and (3) "Active Promotion of Global Environmental

Conservation through International Cooperation" (Article 5). Specific ways to achieve these are: (1) the Basic Environment Plan, (2) an EIA system, and (3) provisions for economic measures. The Basic Environment Plan was formulated in December 1994, though its lack of concrete substance is a shortcoming. In June 1997 the EIA system (see below) finally became law, although it has many limitations.

There has also been some progress on global warming. In December 1997 the Third Conference of the Parties to the UN Framework Convention on Climate Change (COP3) was held in Kyoto, where the Kyoto Protocol was adopted. In response, Japan's government inaugurated greenhouse gas reduction measures in order to fulfill the obligations made at Kyoto. In 1998 the Law Concerning the Rational Use of Energy (Energy Conservation Law) was amended to beef up energy conservation policy, and the Law on Promoting Measures to Remedy Global Warming was enacted to facilitate action on greenhouse gas reductions by the national government, local governments, industry, and the citizens. Nevertheless, as measures based on these laws are woefully inadequate to attain Japan's pledged numerical targets for greenhouse gas cuts, they are merely a first step.

4. Japan's Environmental Policy: Characteristics and Lessons

As the foregoing brief survey shows, Japan's environmental policy evolved as a product of environmental problems and the development of environmental laws to cope with them. Some important characteristics are, first, local government initiatives have played a major role; second, people made aggressive use of lawsuits, which served to leverage advances in policy; and third, serious health damage sustained by so many people led to the creation of an internationally unique system to help victims.

Here we shall expand on these characteristics and see what lessons they have to offer.

4-1 Local Government Initiatives

• Pollution Control Agreements

As pollution steadily worsened in the 1960s, it was the emergence of reform governments in municipalities that effected political solutions. The first of these was the Asukata administration in Yokohama, which was followed by others including Minobe in Tokyo, Kuroda in Osaka Prefecture, and Ito in Kawasaki City. These reform administrations used the authority of self-governance guaranteed in the Constitution to impose specific controls on pollution from factories in their locales. One type of control was the pollution control agreement. The first of these, in Yokohama, was an outgrowth of a large citizens' movement from the beginning of the 1960s to control SO_2 in the city's Naka and Isogo wards. It consisted in a 1964 "gentlemen's agreement" between Yokohama and Tokyo Electric, under which the power company accepted controls on the pollution emitted by a

---

[2] Defined as a pleasant environment, this concept moved environmental policy from the realm of simply protection to creation, as in urban planning.

new thermal power plant it would build. This was a landmark agreement because it demonstrated the possibility of controlling pollution from big corporations, something believed impossible until then.

Having established a precedent, Yokohama went on to conclude similar agreements with other corporations. This was followed in 1968 by an agreement between a Tokyo Electric generating plant in Tokyo and the Minobe administration, which had come into power on a pollution control platform. Negotiations over the terms of this agreement were constantly relayed by the media, which substantially heightened interest among Tokyo citizens. Tokyo mobilized its full contingent of technical and economic experts and presented Tokyo Electric with specific proposals, such as cutting sulfur emissions by using low-sulfur Indonesian coal instead of fuel oil. It was during this process, in April 1968, that Tokyo established the Metropolitan Tokyo Pollution Research Institute in order to join specialists throughout the Tokyo government. As in Yokohama, this agreement with Tokyo Electric was followed by agreements with other major air polluters. These agreements in turn spurred the development of technologies such as dust collectors and flue-gas desulfurization hardware. Air pollution was gradually mitigated and Tokyo's skies began to clear.

Tokyo's initiatives in turn influenced neighboring Chiba Prefecture, which had air pollution from its heavy and chemical industry complexes. Chiba's agreement with a local Tokyo Electric thermal generating plant led to similar agreements with complexes and companies in other parts of the country, so that by 1974, 1,292 local governments in 40 of Japan's 47 prefectures had concluded some sort of pollution control agreement. While these agreements by local governments did achieve results, they also suffered from problems and limitations. For instance, as they were gentlemen's agreements, how could corporations be forced to submit in case of noncompliance, and how would local governments monitor corporate actions to assure agreements were being carried out? One answer was the pollution control ordinance.

• **Pollution Control Ordinances**

Article 94 of Japan's postwar Constitution says, "Local public entities shall have the right to... enact their own regulations within the law," which the Local Government Act carries over by authorizing local governments to establish ordinances. In July 1969 Tokyo was the first to use these provisions to establish the Metropolitan Tokyo Pollution Control Ordinance, which was a more full-blown ordinance than the one it superseded. It included provisions for a system to authorize the building of factories, limitations on factory siting, and the right to request that the supply of industrial water be stopped. Subsequently added were highly significant provisions including the establishment of a Pollution Monitoring Committee that guaranteed broad citizen participation, and an obligation that the governor release information to the public. Further, the publication of *Pollution and Tokyo* by the Metropolitan Tokyo Pollution Research Institute in March 1970 had a profound impact on local gov-

ernments around the nation.

In light of these events, it is safe to say that Tokyo's ordinance and other such local government initiatives constituted a breakthrough leading to a solution for the pollution problem, which illustrates the immense significance of Japan's postwar institution of self-government in the development of environmental policy.

4-2 Environmental Lawsuits

• **The Four Great Pollution Lawsuits**

Environmental lawsuits also played an important role. Of particular note are the four representative postwar lawsuits: two for Minamata disease, and one each for itai-itai disease and Yokkaichi pollution. Although these incidents' causes were established, the companies responsible paid no compensation to the victims, who therefore resorted to lawsuits. Despite the difficulties, plaintiffs won victories in all these lawsuits owing to the many lawyers who feel that achieving social justice is their mission, and to the valuable court testimony by researchers. The Toyama District Court and Nagoya High Court (itai-itai disease, 1971 and 1972), Niigata District Court (Niigata Minamata disease, 1971), Tsu District Court (Yokkaichi pollution, 1972), and Kumamoto District Court (Kumamoto Minamata disease, 1973) all recognized that the defendant companies were totally responsible. Another fact of great significance is that in the course of these lawsuits several legal theories were established, including *causality* based on epidemiological cause and effect; *negligence* relating to the obligation for prediction that in turn assumes the obligation for research and studies, and of the obligation to avoid pollution consequences; and *joint torts*, which imposes joint liability on two or more businesses. The decisions in these four lawsuits had a major impact on concrete advances in subsequent environment-related government administration. For example, they induced inclusion of the absolute liability rule in the Air Pollution Control Law and the Water Pollution Control Law. And by recognizing the joint torts liability of two or more companies that are air pollution sources, the Yokkaichi lawsuit expedited the creation of Japan's Health Compensation Law.

• **Pollution Injunction Lawsuits**

Although the Four Great Pollution Lawsuits all ended in victory for the plaintiffs, people saw that it was more important to prevent pollution than to find redress for victims, which led to the filing of lawsuits seeking injunctions to stop actions causing pollution and environmental damage.

The Osaka International Airport and Nagoya Shinkansen lawsuits, both of which involved noise pollution, sought payments for damages as well as injunctions based on personal and environmental rights. In Osaka plaintiffs wanted to prohibit night flights, and in Nagoya they wanted train noise kept under a certain level. Environmental rights lawsuits were filed in a number of places seeking the prevention of environmental damage. But unlike suits demanding compensation, those seeking injunctions did not always proceed according to plaintiffs' wishes. The Osaka case, for instance, was fought all the way to the Supreme Court's Grand Bench, which on December 16, 1981 ruled that the injunction demand was not lawful for a civil suit, while upholding the original decision that recognized the demand for compensation. This case later ended in a compromise. The Nagoya case went to a second trial, and on April 12, 1985 the Nagoya High Court turned down the request for an injunction on the grounds that noise in the area under litigation did not exceed the endurable level, but it did recognize compensation. This case too subsequently ended in a compromise.

• Environmental Rights Lawsuits

There were also lawsuits seeking injunctions on the basis of environmental rights, whose grounds are found in Articles 13 and 25 of Japan's Constitution.[3] Plaintiffs argued that violation of these rights allowed them to demand injunctions, but the courts dismissed nearly all these cases on grounds such as: (1) The provisions of Articles 13 and 25 outline general principles, and do not make it possible to directly find any specific right of claim; (2) with respect to environmental rights, it is not clear who the legal entities are, what extent of the environment is covered, or what extent of environmental rights violation would call for an injunction; and (3) assuming it is possible to demand injunctions at the stage where environmental damage exceeds the individual's benefit and protection of the law, grounds in positive law would be required. Courts have continued to deny environmental rights. Currently in Japan personal rights have more or less been established, but courts still do not recognize environmental rights.

• Air Pollution Lawsuits

Beginning in the latter half of the 1970s victims of air pollution filed a series of major lawsuits around the country demanding compensation payments and injunctions to keep SO_2, NO_2, and other pollutants from exceeding EQS. These suits included those in Chiba (against Kawasaki Steel), Nishiyodogawa in Osaka, Kawasaki, Kurashiki, Amagasaki, and south Nagoya. The significance of these lawsuits was that they criticized the substantial relaxation in 1978 of the NO_2 EQS and the total cancellation of class 1 regions designated under the Health Compensation Law for air pollution, and they tried to recover the losses not covered by benefits under the Health Compensation Law. Many of these lawsuits have already concluded and generally recognize plaintiff claims for compensation, but the courts have turned down demands for injunctions on the grounds that they are not lawful because when making such abstract nonfeasance claims as "keep air pollution below the EQS," the substance of a claim is unspecified. Lawsuit results are as follows. In the Kawasaki Steel lawsuit in Chiba, plaintiffs won their demand for compensation in the first hearing, then compro-

mised in the second. In Nishiyodogawa, Kawasaki, and Kurashiki, courts recognized the joint torts of multiple companies, while in Nishiyodogawa and Kawasaki claims against the defendants, i.e., the authorities who built and managed the roads, were rejected on the grounds that causality between NO_2 and health impacts was not demonstrable. Nevertheless, in the subsequent second to fourth Nishiyodogawa lawsuits the court recognized the liability of road management authorities for the first time. Specifically, it recognized that air pollution in roadside areas had in the past been heavy. Further, in the second through fourth Kawasaki lawsuits the court concluded that current roadside area air pollution is a cause of health damage, and recognized the liability of road management authorities. By the end of 1996 compromise settlements were also reached between the plaintiffs and the defendant companies in the Nishiyodogawa and Kawasaki lawsuits, and in 1999 in the Kurashiki lawsuit. These settlements attracted great interest as they included the new element of compromise payments meant to rejuvenate the affected geographical areas. In the Nishiyodogawa and Kawasaki lawsuits against road management authorities, plaintiffs reached compromise settlements with defendants because the former were favorably impressed with the latter's efforts to improve roadside environments.

These examples illustrate a major characteristic in which the progress in environmental lawsuits heavily influenced the concrete form of environmental policy.

4-3 Significance of the Health Compensation Law

• Pollution-Related Health Damage Compensation Program

Japan was faced with the imperative task of helping the victims of health damage caused by severe pollution beginning in the 1960s, and one of the unique ways of doing this was for local governments in places with serious health damage caused by air pollution to shoulder part of the self-paid portion of victims' health insurance. But as such measures were at best partial solutions, and because not that many local governments moved to help those in need, victims had no choice but to file civil lawsuits against the polluters in order to obtain full redress, leading to the aforementioned lawsuits.

Some kind of system was needed at the national level, and it was the 1967 Basic Law for Environmental Pollution Control that finally provided for such a system. In response to this, the Law on Special Measures Concerning Redress for Pollution-Related Health Damage was established in December 1969. The idea at that time was to help victims of pollution-induced health damage by instituting an absolute liability system that would be part of civil compensation, but it was actually institutionalized as an administrative redress measure. Under this law payments made to victims were limited to costs for their medical treatment, with one-half of those costs covered by industry, and the other half by the national and local governments, but there was a great deal of criticism for reasons including the limitations

---

[3]   Rights to life, liberty, and the pursuit of happiness; rights to wholesome and cultured living.

on eligible victims and the help they could receive, and the vague principles by which polluters paid costs. Subsequently the government established laws partially revising the Air Pollution Control Law and the Water Pollution Control Law, but victims were still obliged to seek redress by demonstrating civil liability, so there was still the task of creating a real redress program, and the situation changed significantly in July 1972 when the decision in the Yokkaichi pollution lawsuit totally recognized the liability of six industrial complex companies to pay compensation for health damage caused by air pollution. This change occurred because industry came to see the need to set up a relief program, of which it had been disapproving until then, because it feared that many such lawsuits would happen around the country, and because it hoped the program would head off such lawsuits. This situation underlay the October 1973 promulgation of the Health Compensation Law, which superseded the Law on Special Measures Concerning Redress for Pollution-Related Health Damage and took effect on September 1, 1974.

• Later Developments and Problems

Thus it was that in September 1974 Japan created a program that was quite unique, even internationally. Yet, that program underwent a major revision in 1987, and in March 1988 the government canceled all the 41 areas of the country that had been designated as class 1 regions, which stopped redress and compensation for new patient certifications (compensation benefits continued for already certified patients, who numbered 82,830 as of 1996). It was an October 1986 recommendation by the Central Council for Environmental Pollution Control that induced the total cancellation of area designations, an action that drew stiff criticism from air pollution experts and others. The recommendation stated, "We can no longer say that air pollution at present is primarily responsible for bronchial asthma and other disorders," but there are undeniable health impacts of some kind from other substances such as NOx, particulates, and diesel exhaust, which are now the main pollutants. Since then projects to prevent health damage have been scheduled, mainly in the canceled designation areas.

5. What the Future Holds

5-1 The 1990s

We have already covered the 1993 Basic Environment Law and the Basic Environment Plan, which were part of new headway in environmental policy after a period of stagnation. Here we would like to briefly mention some other events of the 1990s.

• Environmental Ordinances by Local Governments

The local government initiatives that had been so important in the formation of Japan's environmental policy from the second half of the 1960s through the first half of the 1970s gradually faded with the political decline of the reformist local governments from the second half of the 1970s and into the 1980s. Critical factors behind this were the worsening regional financial crises around the country, and the "regional administrative reform" it triggered, which consisted in regressive trends including cuts in citizen welfare and the environment. Especially the 1980s were a time when local government and environmental administration suffered great setbacks, but with the arrival of the 1990s the national government passed the Basic Environment Law, to which local governments responded with new basic environment ordinances. Many of them, however, have only abstract provisions modeled on the national Basic Law and almost totally lack autonomy, thus failing to be effective tools for accommodating each locale's unique challenges and requirements. Of course some local governments did incorporate measures for autonomy as in Kawasaki City's ordinance, which included provisions for environmental rights and stipulations for planning assessments, or Shiga Prefecture's ordinance, which established an Environmental Self-Government Committee system in order to promote citizen participation. It remains to be seen how much these local government initiatives will do to usher in another new era of environmental protection.

• Environmental Impact Assessment Law

Since the latter half of the 1970s Asian countries have been gradually institutionalizing environmental impact assessments, but progress toward an assessment law in Japan was obstructed by opposition from industry and business/ project agencies. The first EIA assessment was finally submitted to the Diet in 1981, but died in 1983 without deliberation. Attempts at legislation having been abandoned for the time being, EIAs were conducted as so-called "Cabinet assessments" under the "Environmental Impact Assessment Implementation Scheme" approved by the Cabinet in 1984. Owing to the lack of a national assessment law, local governments began in the early 1970s to set up their own assessment systems, called by contrast "local government assessments." Fukuoka Prefecture was the first, in 1973, to develop a system, followed by a Kawasaki City ordinance in 1976. As of April 1997, seven local governments carry out EIAs according to ordinances, and 44 on the basis of other systems. Not a few of these local government assessments are more advanced than national government Cabinet assessments because some require public hearings, or the establishment of third-party organizations to increase the reliability of investigations (Kawasaki City, for example, has an Environmental Impact Assessment Deliberative Committee).

Cabinet EIAs, moreover, have drawn much fire owing to a number of limitations such as (1) no requirement to consider alternatives, (2) insufficient participation by concerned individuals, and (3) results of Cabinet EIAs cannot be fully reflected in project licensing. And encouragement of EIAs by Article 20 of the Basic Environment Law helped heighten the national mood toward legislating EIAs, which finally came about in June 1997, making Japan the last OECD country to pass such a law.

This Environmental Impact Assessment Law can be characterized as follows in comparison with the Cabinet assessments. First, it covers more project types. (1) Newly subject to EIAs are existing railways, large-scale forest roads,[4] and the like, and (2) in addition to projects always requiring EIAs (class I projects), the law stipulated a screening process for class II projects, whose necessity for EIAs is determined on a case-by-case basis. Electric generating plants, a point of contention until the very last, were also made subject to assessments. Second, while Cabinet assessments at the outset listed in their technical guidelines the items to be studied, the law introduced scoping in order to clarify the points at issue and carry out the assessment efficiently. This involved narrowing down to certain items by considering the characteristics of the locale and its environment. Third, the law expanded participation by concerned parties. Specifically, (1) there are two opportunities to submit opinions instead of one (at the scoping and draft stage), and (2) anyone can submit an opinion because submitters are no longer required to be local residents. Fourth, the Environment Agency director-general could formerly state an opinion only when requested by the minister overseeing a project, but the new law makes it possible to state opinions when necessary. Fifth, the law requires parties granting project licenses and permits to determine if proper attention is accorded to conserving the environment as written into assessment documents to ensure that EIA results are reflected in project licenses and the like. Sixth, there were previously no arrangements for EIA followups, but the law has provisions for followup studies, and it is possible for contractors to redo EIAs for projects unstarted for a long time.

As this shows, the Environmental Impact Assessment Law represents a measure of progress over the earlier Cabinet assessments, but problems yet remain. First, the law does not expressly call for the consideration of alternatives, which is the core of EIAs. Second, in recent years the international focus is on conducting EIAs (strategic environmental assessments, or SEAs) at an earlier stage such as when formulating policy or at the preliminary planning stage, but this provision was considered premature and excluded. Third, the law does not cover overseas or official development assistance (ODA) projects, so implementation is left to the contractor's voluntary efforts and to guidelines. And fourth, although there is indeed greater opportunity for the environmentally concerned to participate, the law does not allow participation after screening and the submission of opinions, and public hearings are not required. Other substantial improvements needed include (1) establishment of a neutral screening authority, (2) enhancing means of correcting flaws in EIAs, and (3) participation-supporting institutions (such as information centers and funding).

---

[4] "Forest road" in Japan means any kind of road through forested areas, from logging trails to paved highways.

5-2 International and Global Responses

Since the beginning of the 1990s Japan too faces the crucial challenge of how to deal with international and global environmental problems. This section shall briefly discuss Japan's responses in that area.

• International Environmental Problems

These arose when existing pollution and environmental problems began to spread internationally. They include: (1) Transboundary, broad-area pollution, a typical example of which would be damage from acid rain; (2) so-called "pollution export" (particularly serious in this respect is the transfrontier shipment of hazardous wastes); (3) problems associated with international trade and transactions, such as the mechanism and rules of current "free trade," which accelerate the plundering devastation of environments and resources, especially in the developing countries, and problems concerned with international transactions in wild flora and fauna controlled by the Convention on International Trade in Endangered Species of Flora and Fauna (CITES); and (4) problems associated with ODA and other assistance to developing countries (in this respect there has been heightening international criticism of projects such as large dams, which lead to serious environmental damage). Japan faces the vital task of setting up systems and new rules to deal with these problems internationally. Moreover, many of the problems noted here are closely connected with the overseas presence of Japanese corporations and the internationalization of the Japanese economy itself, both of which have progressed rapidly since the 1980s. Since the 1990s, therefore, Japan's environmental policy has been called into serious question not only as to its response to domestic problems, but its international implementation as well, and especially in Asia. However, the almost total lack of an active response so far makes this a vital task for the future.

• Global Environmental Problems

Generally the international problems described above are included in global environmental problems, but in a strict sense the latter are problems in geographical areas or problems of a nature not subject to national sovereignty.

First, problems in areas not subject to national sovereignty include the open seas, deep sea beds, the ozone layer, Antarctica, and space, and involve pollution prevention, conservation of resources, development/use, management, and other issues in those places. Second are problems concerning the global commons, which include the conservation of climate systems, and nature or biodiversity. International regulation of conservation, use, and management is an especially vital issue. In Japan too the response to such global environmental problems has been called into question, and a great deal is left to be desired in terms of concrete initiatives and new legislation in related areas.

Let us begin by pointing out a number of problems involving international public areas. Conventions such as the United Nations Convention on the Law of the Sea and the

International Convention for the Prevention of Pollution from Ships (MARPOL) require their signatories to protect the environment in international waters, but Japan has not done enough in this respect. In 1996, before ratifying the Law of the Sea Convention, Japan passed corresponding domestic legislation, and in response to the MARPOL requirement for double hulls, Japan required them for new ships ordered after July 6, 1993 and for large tankers over 25 years old under a ministerial ordinance provided for by the Law Concerning the Prevention of Marine Pollution and Marine Disasters. Further, there is work on cooperation for port entry and immigration in the Asia-Pacific region, and Japan has adopted a Northwest Pacific Regional Marine Action Plan covering the Sea of Japan and the Yellow Sea. But these regional sea arrangements are still just basic frameworks that need to be fleshed out. In order to conserve living resources in international waters, Japan prohibited driftnetting by Japanese vessels in the North and South Pacific, but this prohibition does not provide distinct criteria and measures for the management of living resources on the open seas in general.

In order to domestically accommodate the Montreal Protocol, Japan in 1988 passed the Ozone Layer Protection Law, which enacts measures to limit the production and consumption of controlled substances and to rationalize their use. One problem here, however, is that even though the harmfulness of HFCs, HCFCs, and other CFC substitutes has been noted internationally, Japan has instead pursued the switch to those substances. Japan also established the "Conference to Promote Ozone Layer Protection" in order to promote the recovery, reuse, and destruction of controlled ozone depleting substances. It has created measures to encourage this, and has asked local governments and the related industries to implement them, but the recovery rate is still extremely low. A system under which consumers would foot the bill is under consideration, but further enhancement of measures is an urgent priority. There is also hardly any progress in reducing use of the soil fumigant methyl bromide.

The 1991 Protocol on Environmental Protection to the Antarctic Treaty (Madrid Protocol) sets forth broad obligations including EIAs when planning activities in the Antarctic region, conservation of flora and fauna, and management of wastes, and Japan in 1997 passed the Law Concerning Protection of the Environment in the Antarctic Region, which corresponds to the Madrid Protocol. Under this law, plans for all activities in the Antarctic region require applications in advance, and the Environment Agency director-general must make sure that there are no actions prohibited under the protocol, and that the Antarctic environment is not seriously affected. There are also limitations on mineral resource activities and collection of plants and animals. The law also calls for edification in this regard for Japanese citizens who visit Antarctica as tourists.

Second is the global commons issue. Japan has accommodated the Framework Convention on Climate Change by taking action in accordance with the Action Program to Arrest Global Warming, which was adopted in 1990 by the Council of Ministers for Global Environmental Conservation. For the time being the Action Plan's objectives are stabilizing per capita and total CO_2 emissions at their 1990 levels in 2000 and beyond. In 1979 the Energy Conservation Law had been established, providing low-interest financing and tax breaks for installing energy-saving equipment, and promoting new energy sources such as geothermal, photovoltaic, and wind in addition to nuclear power. In 1998 Japan's government, in a bid to discharge its obligations as set forth in the Kyoto Protocol, established the Blueprint for Measures to Counteract Global Warming, amended the Energy Conservation Law, and passed the Law on Promoting Measures to Remedy Global Warming. Despite these measures, however, attainment of Japan's GHG reduction goals is as distant as ever. In the transport sector, for example, Japan should in this respect be making serious efforts such as switching to public railways and coastal shipping, introducing low-emission vehicles, and building an efficient distribution system. The government is also trying to reduce wastes and use the waste heat from incineration plants and sewage, and calling on the public to save energy, stop their car engines when parked, and reduce household wastes and emissions through environmental accounting, but industry is still dragging its feet. Japan also needs aggressive measures that include not only legal controls, but economic incentives as well. And because this problem spans various sectors, it will also be necessary to have laws such as a "Basic Law to Prevent Global Warming" that would make comprehensive measures possible.

In 1995 the Environment Agency's Nature Conservation Bureau drew up a "Vision for Nature Conservation" and, on the basis of the Convention on Biological Diversity, a "National Strategy for Biological Diversity," but formulation of the latter had little input from NGOs and citizens. The Strategy's specific provisions are not based on fundamental policy, and it merely sets forth existing measures exercised under fractionalized administrative procedures. It is not comprehensive, well planned, or effective.

6. Summation

The foregoing has been a brief discussion of pollution and environmental problems in postwar Japan, and the development of policy to deal with those problems. Henceforth Japan will have to subject its experiences to an historical examination once again, and address both its remaining domestic problems and the various new problems that have cropped up. In conjunction with this, Japan must in particular gain a self-awareness of its own place in Asia and discharge its international responsibilities and role.

(MIYAMOTO Ken'ichi, AWAJI Takehisa, SHIBATA Tokue, ISONO Yayoi, TERANISHI Shun'ichi, OKUBO Noriko, ISOZAKI Hiroji)

Essay 1   A Perspective on International Comparative Analyses of Environmental Problems

Generally when an environmental problem crops up in a certain country, we must consider the underlying political and economic factors, which can be categorized roughly as follows. (1) Capital formation in the public and private sectors (especially capital spending for pollution control and the building of social capital for environmental conservation such as housing, sewerage, and parks); (2) industrial structure (proportion of heavy and chemical industries, which generate a heavy pollutant load; management and recycling of industrial wastes, etc.); (3) regional structure (urbanization, especially the maldistribution of economic capacity and population, as in the concentration of everything in Tokyo); (4) transportation system (especially the state of railways and other public mass transit systems, and the proportions of motor vehicle use in transporting people and freight, etc.); (5) consumer lifestyle (consumption of energy and other resources, and waste generation, by mass-consumption, mass-disposal lifestyles; waste recycling, etc.); (6) public inter-

vention; involved here are: (a) basic human rights (not only the right to exist, but also the legislation of social rights, environmental rights, and the like, and the extent to which they are translated into everyday life and social movements); (b) mass media, public opinion, and social movements (the extent to which society guarantees the rights of free thought, public discourse, academic research, expression, and association); and (c) democracy (the extent to which separation of powers is guaranteed, especially the independence of local self-government and the judicature); and (7) internationalization (the way in which government, business, and NGOs are working on international environmental policy).

By studying the foregoing seven factors and conditions for each country, we can perform an international comparative analysis and examination on the causes of environmental problems and the environmental policy challenges that arise in dealing with them.

(MIYAMOTO Ken'ichi)

Essay 2   Pollution-Related Health Damage Compensation Law: How the System Works

The Health Compensation Law system works as follows. Regions are classified and designated as class 1 (air pollution-related nonspecific illness) or class 2 (specific illness) depending on whether there is a specific relationship between the substance causing pollution-related health damage and the illness. Regions designated class 1 are defined as "regions where marked air pollution has arisen over a considerable area as a result of business activity or other human activities, and where diseases... due to the effects of such air pollution are prevalent." Class 2 regions are the five designated regions for Minamata disease, itai-itai disease, and chronic arsenic intoxication. Compensation payments are made to people who satisfy the three requirements of (1) designated illness, (2) designated region, and (3) exposure, and who apply for screening and are certified as patients through a screening by the governor of the prefecture (or the mayor of the ordinance-designated city) in which the designated region is located. Certification is granted on the basis of reports by a certification council created by the governor or mayor.

The intent is to provide quick relief and compensation, but in the case of Kumamoto Minamata disease, for example, procedures were seriously delayed, leaving many applicants uncertified for years, which led to the filing of lawsuits to determine the illegality of nonfeasance. Along with the re-

gional designation, certification criteria are also a crucial determinant for receiving compensation. There are seven types of compensation benefits: medical care expenses, medical care allowance, funeral expenses, handicap compensation, compensation for survivors, child compensation allowance, and lump compensation payment for survivors, which are all limited to expenses incurred since the time application is made. In other words, past expenses being not included, part of the liability for compensating damages is institutionalized. And while compensation payments in class 1 regions are shouldered by both fixed and mobile pollution sources, their proportions are determined by Cabinet order. In the case of mobile sources, it is covered by a reserve created from automobile tonnage tax, while for fixed sources, in each Cabinet-designated region, each factory or other place of business emitting at least a certain maximum quantity of gases pays a monetary amount (pollution load levy) calculated by multiplying the Cabinet-determined load factor for each region by the Cabinet-determined emission quantity of health-impacting substance (SOx). If the system were in line with the principle of civil liability toward certified victims in the original sense, this levy would properly be calculated from past SOx emissions causing illness, but it is actually calculated from the previous year's emissions. The levy is also imposed on emission sources outside of designated areas, although at a lower rate.

The Health Compensation Law's purpose is not only paying compensation benefits to make up for health damage, but also to carry out health and welfare projects meant to provide for the recovery, maintenance, and improvement of victims' health, and to prevent health damage by designated illnesses. Half the funds for these projects come from pollution load levies and automobile tonnage tax, and the remainder from the national and local governments.

(ISONO Yayoi)

Chapter 2
Republic of Korea

54

1. Introduction

In the 1960s South Korea embarked on full-blown economic development and achieved a certain measure of progress. In 1995 the country's per capita GDP surpassed $10,000 for the first time, and in the following year it joined the OECD, thereby entering the ranks of the developed countries. Very rapid industrialization and urbanization have occurred in South Korea through this process of economic development.

One characteristic of this process is the disproportionate concentration of everything in Seoul. In that connection, let us briefly examine the changes in South Korea's urban population size. In 1944, just before liberation from Japanese rule, total population on the Korean Peninsula was 15,880,000, with 2,060,000, or a mere 13%, living in the cities. When the plan for economic development was initiated in 1962, South Korea's urban population started to quickly burgeon, so that in 1970 the cities hosted 41.8% of the population. In 1991 South Korea had a total population of 43,270,000, of which 33,020,000, or 76.3%, were urbanites. Accounting for part of this number are villages that were elevated to city status, but there is no mistaking that South Korea's urban population growth has proceeded faster than in postwar Japan. This is also the same process that augmented convergence into Seoul, which according to 1993 data had a population of 10,670,000 (24.2% of national total), 323.7 trillion won in savings (61.5%), 69.8 trillion won in gross regional product (26.1%), and 3.26 trillion won in regional taxes (29.6%). This concentration of everything into Seoul is the most serious regional problem that South Korea faces today, and it also has a significant influence on the occurrence and structure of environmental problems in this country.

Fig. 1 Principal Industrial Parks in South Korea

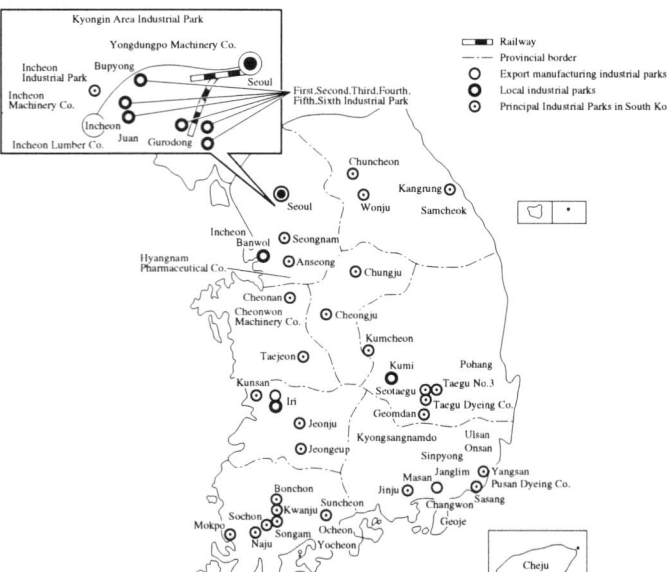

Source: *South Korea's Economy at a Glance*, Korean Industrial and Economic Research Institute, p. 34.

Pollution and other environmental problems became a pressing matter for society mainly with frequent events of industrial pollution at the industrial parks developed around South Korea to accommodate the country's rising heavy and chemical industries beginning in the 1970s. In recent years lifestyle changes especially in the cities have engendered serious urban pollution problems such as managing domestic wastewater and solid wastes, and pollution caused by the growing motor vehicle fleet. In many ways these problems happened in basically the same way they did in postwar Japan, a difference being that they have occurred in a much shorter time span, making them nearly simultaneous.

While keeping differences with Japan in mind, this chapter presents a general overview of environmental problems in South Korea, the development of environmental policy to deal with them, and their characteristics. It also briefly discusses future challenges and the outlook.

2. Pollution and Environmental Problems in South Korea

2-1 Ulsan-Onsan Industrial Complex Pollution

It was in the 1960s, after South Korean economic development began, that people became aware of environmental problems. From that decade and through the 1970s such problems were manifested primarily as serious industrial pollution. A representative example would be pollution at the Ulsan and Onsan industrial parks.

- **City Created by Economic Development**

In 1962 Ulsan Industrial Park was the first industrial park designated by the South Korean government for fostering the heavy and chemical industries, and it was representative of that country's economic development in the 1960s and 70s. Ulsan City was designated South Korea's first wide-area city[1] on July 15, 1997, and on that occasion Ulsan County with its neighboring Onsan Industrial Park was merged with Ulsan City, which had been promoted to a city in 1962 by agglomerating the surrounding area. Though the population was only about 80,000 then, it is now a large city of 1 million and literally a product of economic development. There are actually two industrial parks here — Ulsan Mipo Industrial Park and Ulsan Industrial Park — but this chapter will refer to them as Ulsan Industrial Park (see Fig. 1 for location).

- **Overview of Ulsan Industrial Park**

1995 data show that production and exports were, respectively, 29.6 trillion won and $13.7 billion for Ulsan Mipo, and 6.9 trillion won and $2.8 billion for Onsan. Ulsan Mipo

___

[1] Wide-area cities, of which there are currently six, were created when cities directly governed by the Interior Ministry until that time were merged with parts of their surrounding rural areas.

had 523 factories with at least 10 employees each, and a total of 110,049 workers, while Onsan had 97 such factories and a total of 11,375 workers. As the amount of products shipped by South Korea's manufacturing industries as a whole in 1993 was 128 trillion won, Ulsan Industrial Park accounts for a high proportion of the manufacturing sector. In 1980 it accounted for nearly 49%, but in recent years that proportion has gradually declined.

• Air Pollution

Air pollution in the Ulsan area gradually became serious beginning around 1967 when the factories here began full-scale operations. It was the worst from the end of the 1970s and into the 1980s, during which time the SOx concentration in Ulsan City never fell below 0.1 ppm, and in 1978 near the factories it hit the shocking level of 2.87 ppm. Recently with the use of low-sulfur fuels there has been gradual improvement in this pollutant, but rapid growth in motor vehicle use is increasing NO_2 emissions. South Korea's motor vehicle fleet was only 129,371 in 1970, but crossed the 10 million mark in July 1997. 1995 government data show that the annual averages of SO_2 and NO_2 are 0.028 ppm and 0.023 ppm, respectively, but citizens and NGOs have little faith in government air pollution statistics.

• Water Pollution

Ulsan Bay was one of South Korea's prime fisheries prior to industrial park development, but now there is hardly any fishing due to the heavy pollution from industrial effluent. There is little progress in government studies of the extent of this water pollution. A survey by Pusan Marine University in 1985, when pollution was said to be at its worst, revealed marine metal concentrations for cadmium, lead, and zinc that were, respectively, 2,000, 3,000, and 400 times higher than the corresponding USEPA standards.

Since 1985 a serious problem in the Ulsan area has been the pollution-induced "Onsan disease." Many of the people living near Onsan Industrial Park had symptoms similar to those of Japan's itai-itai disease, leading to suspicions of pollution disease, but the situation has yet to be sufficiently elucidated (see Part I Chapter 3 for details). With pollution illnesses the most important thing is to determine how people are being harmed, but Ulsan City authorities take the official position that there is no pollution-induced harm because both air and water pollution in the area are within the environmental standards.

• Relocation Hides the Problem

Of important note is a large-scale program by the central government to relocate people harmed by the pollution in the Ulsan area. Initially the program was meant to relocate somewhat over 10,000 people (Table 1), but later relocated about 8,400 households totaling 37,600 people. Its basic problems are, first, the program was definitely not aimed at solving the pollution problem itself, but instead at creating an environment where the factories could continue their operations without worrying about effects on people. Second, and graver, is that relocating area residents made it very difficult to find out what Onsan sickness is, which was tantamount to papering over the problem.

Since that time pollution in the Ulsan area has not been limited to the air, water, and noise pollution caused by the industrial parks, for added to this are the water pollution by household graywater, motor vehicle air pollution, and other new city-type pollution caused by rapid urbanization, thereby creating a situation that is representative of the current state of pollution throughout South Korea.

2-2 New Types of Pollution from Urbanization

In addition to environmental problems such as those described above, South Korea is seeing the appearance of a new set of problems in conjunction with urbanization.

To begin with, advancing urbanization and rising incomes are inducing great change in traditional South Korean lifestyles, so that problems similar to those faced by Japan beginning in the 1970s are now increasing in gravity. With water pollution, for example, there is now urban graywater in addition to preexisting factory effluent, because of which water quality in all South Korea's major rivers is steadily worsening. According to the *1996 Environment White Paper* by South Korea's Environment Ministry, biological oxygen demand (BOD) readings in the country's four large rivers are 1.0-4.4 mg/l in the Han River, 1.2-7.3 in the Nakton River, 1.2-4.8 in the Kim River, and 1.5-7.0 in the Yeongsan River. Water quality in their watersheds does not even meet the class 3 standard, which can stand up to industrial use. On the Nakton River this has led to a regional dispute between Taegu City on the upper reaches and Pusan City on the downstream portion. This river also suffered a major phenol spill in 1991.

In recent years South Korea is having exactly the same grave problem as Japan with the disposal of wastes. South

Table 1 Occupations of Citizens Subject to Relocation

(Numbers of individuals)

	Agriculture	Fishing	Industry	Other	Uncategorizable	Totals
Ulsan	1,026	363	1,505	4,176	1,000	8,070
Onsan	492	461	-	1,153	-	2,106
(Coastal)	293	461	-	645	-	1,399
(Inland)	199	-	-	508	-	707
Totals	1,518	824	1,505	5,329	1,000	10,176

Note: This was commissioned to Seoul National University by the South Korean Environment Agency.

Source: *Study on Relocation of People Harmed by Ulsan/Onsan Industrial Park Pollution — Final Report*, Nov. 1984, p. 783.

Korea roughly classifies wastes as " household waste" and "business wastes" (the latter being further broken down into "factory wastes" and "designated wastes"), and according to 1994 data total waste generated came to 147,049 tons per day, the breakdown being 58,118 tons (39.5%) for household waste and 88,931 tons (60.5%) for business waste. On April 1, 1995 (one year earlier in some areas) the household waste disposal fee system changed from a fixed fee assessed on the basis of each household's property tax and building area to a quantity charge according to the amount of a household's wastes. Thanks to this switch, the amount of wastes generated apparently dropped about 27% in the first year, but subsequently increased again (see Essay 1 at the end of this chapter).

A bigger problem is wastes classified as "designated wastes," whose categories are waste acids, waste alkalis, waste oil, waste organic solvents, waste synthetic resins, waste synthetic rubber, dust, sludge, and other, meaning that "designated wastes" are hazardous industrial wastes. Still, the amount generated in 1994 was 1,351 tons per day, which was a mere 0.9% of total wastes generated. This amount is only 0.1% of industrial wastes generated in Japan, which come to about 1 million tons a day. Note, however, that the types of wastes covered by South Korea's "designated wastes" category differ. For example, this category does not include construction and demolition wastes or livestock manure, which in 1992 accounted for 16% and 19%, respectively, of industrial wastes in Japan.

South Korea has basically disposed of wastes by landfilling, but in recent years there has been a critical shortage of landfill space, which is leading to increasingly antagonistic disputes on the outskirts of Seoul and throughout the country, and creating the same situation as in Japan. Data from South Korea's Environment Ministry show that, in 1994, 53% of wastes were landfilled, 4.3% were incinerated, and the remainder were recycled. These figures suggest that the recycling rate is over 40%, but in contrast with Japan's low 3.9% recycling rate for municipal wastes in 1992, this is an incredibly high figure,[2] and requires more detailed study.

3. Characteristics and Problems of Environmental Policy

Below is a survey of South Korea's environmental policy and how it has evolved, and a discussion of its characteristics and problems.

3-1 Environmental Law

Ro Zaishik, who has long been involved in the development of South Korea's environmental policy, observes that

---

[2]   The rate for South Korea comes from the *1996 Environment White Paper* by South Korea's Environment Ministry, while that for Japan is from the Environment Agency's *Quality of the Environment in Japan*, 1996 edition.

environmental law in this country has evolved through three stages. First is the stage from 1962, when economic development planning started, to 1977. The 1963 Pollution Control Law is a product of those years, but this was also a time when economic growth was a national priority. The second stage was from 1978 to 1987, during which time the Pollution Control Law was abolished and replaced by the Environmental Conservation Law and Marine Pollution Prevention Law (end of 1977). In October 1978 the government released the Nature Protection Charter. Although this was when people finally became aware of the need for compatibility between economic growth and conservation, the country was still oriented mainly toward the former. Nevertheless, some progress was achieved, such as the 1980 creation of South Korea's Environment Agency (under the Health and Society Ministry) as the country's first central government organization specializing in the environment, and the addition to the Constitution of an article on environmental rights. In the third stage, which started in 1988, South Korea finally began the full-blown shift toward primary orientation on conserving the environment. During this stage the Environment Agency was promoted to the Environment Bureau (whose administrator was a member of the State Council), and in December 1994 to the Environment Ministry. In 1990 the Environmental Conservation Law was replaced by the Basic Environmental Policy Law, and in conjunction with this came a number of other laws including the Law on Pollution Damage Dispute Negotiation, the Air Quality Preservation Law, and the Water Quality Preservation Law, which achieved groundbreaking progress in South Korea's environmental legal institutions.

3-2 Development Dictatorship and Limitations on Democratic Institutions

Generally, in examining the evolution of a country's environmental policy it is insufficient to just trace the basic progress in its environmental laws. To see how effective they have actually been, one must consider in totality the circumstances relating to political, economic, and social factors manifested in policy. One must pay attention to trends at not only the central government level, but also the local government level. In the 1960s and 1970s in Japan, for example, the rise in citizens' anti-pollution movements, pollution lawsuits, and the support of these by public opinion all played important roles. As described in the previous chapter, local government initiatives such as pollution ordinances and other efforts preceding those of the central government heavily influenced development of environmental policy. By contrast, in South Korea the many years of severe political restrictions placed on local government systems, which are a vital part of democratic institutions in modern states, mean that South Korea has hardly anything to show in the way of environmental policy initiatives on the local government level.

Nevertheless, in March 1991 South Korea held local assembly elections for the first time in 30 years, and in June

1995 there were elections for local heads of government in conjunction with new local assembly elections. These events herald a new age in local government that will no doubt bring forth unique developments in environmental policy on the local government level.

Until now, however, South Korea has been under the control of a development dictatorship run by an overwhelmingly centralized political system. Owing to worsening industrial pollution especially since the 1970s even the central government gradually became aware of the importance of environmental policy, which led to the Environmental Conservation Law in late 1977, although it left much to be desired in terms of effectiveness. The reason was to be found in the country's characteristic political and economic structure of the time, i.e., since 1948 South Koreans have suffered the onus of separation from their brethren in the North, and under those circumstances maintaining and stabilizing a dictatorial system of rule built on rigid military confrontation was the political agenda that had priority over everything — including the environment. South Koreans have therefore lived under political conditions characterized by the imposition of harsh restraints on basic human rights and the institutions of modern democracy, including personal liberties. Especially in the case of South Korea, this has strictly limited the possibilities for social solutions to the serious environmental and pollution damage occurring behind the scenes under the high economic growth that has been considered a model for NIEs-type economic growth in East Asia.

3-3 Environmental Policy up to the 1970s

• Ineffective Environmental Policy

South Korea's 1963 Pollution Control Law, which was based in part on a 1960 Osaka City ordinance, was a highly inadequate law that lacked teeth, and it was not until 1969 that the government passed its enforcement order. Ro Yunghui, the first dean of the Graduate School of Environmental Studies at Seoul National University, said, "This law concerns itself with industrial pollution alone, but not with nature damage or motor vehicle exhaust. Controls cover only air, water, noise, and vibration. There is no total pollutant load regulation and its administrative penalties are very weak. It was known more as the Pollution Permission Law than as the Pollution Control Law." In 1967 a pollution office was established in the Health and Society Ministry. Through this and other actions South Korea gradually built the administrative organization responsible for pollution issues, but personnel and budget shortages meant that it was actually impossible to deal adequately with environmental problems. What is more, the First Comprehensive National Development Plan, which began in 1972, devoted but one page to pollution and environmental problems, making only the following statement: "In the process of rapid industrialization it is expected that henceforth there will be still more pollution. We must take adequate measures in advance to combat pollution instead of repeating the historical errors of

the industrialized countries. As part of these measures it will be necessary to make polluters bear some of the costs."[3] Since 1972 was the year of the United Nations Conference on the Human Environment in Stockholm, South Korea's government was fully aware of the steadily worsening environmental problems in its own country, but it had hardly any specific policy to cope. Especially since the 1970s was a time when the government put its shoulder into implementing its policy for the heavy and chemical industries, the First Comprehensive National Development Plan was very much a social capital construction plan with emphasis on building industrial infrastructure. This resulted in frequent occurrences of industrial pollution around the country from the heavy and chemical industries, but there was hardly any progress in concrete environmental policy until passage of the December 1977 Environmental Conservation Law which, unlike the preceding Pollution Control Law, had a broader perspective on environmental problems including damage to the natural environment, instead of covering only pollutant emissions. Its provisions included establishing environmental quality standards, instituting an environmental impact assessment system, continually measuring the degree of pollution, and setting up a system for contributions to pay pollution control expenses. Other specific laws were passed in conjunction with this law, such as the Marine Pollution Prevention Law and the Law on Synthetic Resin Waste Disposal Operations, and other laws on sewage, poisons, and the like were revised. Under the Environmental Conservation Law South Korea for the first time worked on the minimum legislation needed for the concrete development of environmental policy, but full-blown implementation of that policy had to wait for the collapse of the Pak Jeonghui government, which was the military dictatorship of the day.

• Pollution Victims Stand up for Themselves

Citizen movements against environmental and pollution damage were repressed especially under the Pak government, a situation we shall examine specifically with reference to the case of Ulsan Industrial Park.

Already from around 1967 when many factories began locating in Ulsan Industrial Park, its air pollution was causing real damage as seen in reduced agricultural harvests. Residents of Ulsan City's Nochendong area, where air pollution was worst, immediately organized a committee that started making complaints to the factories, to local administrative offices, and also to the pertinent government agencies. They demanded relocation but the authorities and companies completely ignored them. Later, with the mediation of Kyeongsannamdo Province, industrial park businesses paid a modicum of compensation to farmers and fisherfolk for crop damage and decreased catches, but they refused all other citizen demands. It was not until 1977, when a Pollution Council was established in the Health and Society Ministry, that the central government began considering the re-

---

3 *First Comprehensive National Development Plan*, 1972, p. 82.

location of pollution-victimized citizens, but owing to fiscal restrictions actual relocation did not start until 1985. This was triggered by the appearance in national newspapers of editorials about Onsan disease, which had come to the public's attention that January due to a survey by the Korean Research Institute on Pollution Problems, which had been founded in 1982 as the country's first private environmental organization.

There are several instances in which citizens in Ulsan and other areas won compensation in court for crop damage and decreased fish catches caused by air and water pollution. In South Korea legal resolution of disputes over environmental issues has generally been by civil cases whose demands are limited to compensation for damage to assets such as reduced produce and fish harvests. In South Korea there have been hardly any instances, as in postwar Japan, in which victims filed lawsuits seeking compensation for not only damage to assets, but also damage to health. One exception, however, was the 1989 case in Seoul where people living near a briquette factory sued and won compensation for health damage.

3-4 Developments since the 1980s

It was in the 1980s when developments in South Korean environmental policy finally began to gradually take shape. Environmental quality standards for air pollution were created in response to the founding of the Environment Agency and the addition of environmental rights provisions to the Constitution in 1980. These included those for SO_2 in 1979 and, in 1983, those for NO_2, CO, oxidants, airborne substances, and hydrocarbons. This decade was also a time when other new developments started. For instance, in 1982 the EIA system was implemented, and in 1983 the government instituted an emission levy system. Following is a brief survey of how South Korea's environmental policy evolved.

• Environmental Quality Standards

This section examines the attainment of standards established under the Environmental Conservation Law for the main air pollutants.

The standard for SO_2 was originally an annual average

of under 0.05 ppm, but it was since toughened (Table 2). Because Japan's standards are below 0.04 ppm for the 24-hour average and below 0.1 ppm for the one-hour average, arguably South Korea's standards are still comparatively lax. Still, because the 1996 White Paper reports that the annual average SO_2 concentration is 0.028 ppm, the annual average standard has been met. Compliance notwithstanding, South Korea's air pollution by SO_2 is still very serious. In addition to emissions from factories and business establishments, another significant source of SO_2 emissions is residential floor heating.[4] For example, total 1994 SO_2 emissions were 1,602,764 tons, of which 833,428 tons (51.9%) were industrial, and 164,001 tons (10.2%) were from space heating. And because emissions are largely concentrated in certain areas, SO_2 air pollution is no doubt quite serious in certain places. Such areas need locally stricter environmental quality standards, but the limitations on local government authority described above have prevented this. Under the Environmental Conservation Law, the mayors of certain cities and provincial governors can establish separate environmental standards, but they must have authorization from the central government, specifically the Health and Society Minister, to do so. Providing that authorization is gained, it is not impossible to create separate standards, but that privilege is limited to Seoul, wide-area cities, and provinces. It was therefore legally impossible for Ulsan City, despite its grave air pollution, to establish its own environmental standards until quite recently when it was finally promoted to wide-area city status on July 15, 1997. This situation reveals a serious institutional deficiency in comparison with postwar Japan.

A look at the standards for NO_2, which has subsequently become South Korea's main air pollutant, shows that the standards as listed in Table 2 are laxer than those of Japan, whose 24-hour average standard is 0.04-0.06 ppm or lower (below 0.02 ppm before its relaxation in July 1978). Official Environment Ministry data show that NO_2 concentrations even in South Korea's major cities are below the annual average standard of 0.05 ppm, which puts them in compliance. But here too, real NO_2 air pollution levels give pause

---

[4] A traditional form of heating, called *ondol*, which passes hot air under the floor.

Table 2 South Korea's Atmospheric Environmental Quality Standards

SO_2	Annual avg.	below 0.03 ppm	CO	8-hr avg.	below 9 ppm
	24-hr avg.	below 0.14 ppm		1-hr avg.	below 25 ppm
	1-hr avg.	below 0.25 ppm			
NO_2	Annual avg.	below 0.05 ppm	TSP	Annual avg.	below 150 μg/m³
	24-hr avg.	below 0.08 ppm		24-hr avg.	below 300 μg/m³
	1-hr avg.	below 0.15 ppm	PM-10	Annual avg.	below 80 μg/m³
				24-hr avg.	below 150 μg/m³
Lead	3-mo avg.	below 1.5 μg/m³	Ozone	8-hr avg.	below 0.06 ppm
				1-hr avg.	below 0.1 ppm

Notes   1. TSP: Total suspended particulate matter.
         2. One-hour and 24-hour averages may not be exceeded more than three times annually.
Source: Environment Ministry, *1996 Environment White Paper*, p. 109.

for concern. For example, a survey by a private environmental organization found daily NO_2 average concentrations of 0.048 ppm in Seoul, 0.038 ppm in Pusan, and 0.06 ppm in Ulsan (0.127 ppm in the worst location), values far higher than Environment Ministry figures. In particular, although SO_2 emissions have leveled off recently, NO_2 emissions rise year by year because of increasing vehicular traffic. For instance, total NO_2 emissions increased every year from 878,389 tons in 1991 to 1,191,533 tons in 1994. Sector breakdown is 47.5% for transport and 29.5% for industry, showing that the overwhelming share is from motor vehicles, and requiring the implementation of urgent, comprehensive measures.

A South Korean environmental organization makes the following observations about problems in the way air pollution is assessed. First, the automatic air pollution monitoring network is highly inadequate, with 78 monitoring stations nationwide (increased to about 100 stations presently). By comparison, Japan currently has 1,331 environmental background monitoring stations and 295 motor vehicle exhaust monitoring stations. Second, management and operation of the atmospheric monitoring stations are highly inadequate, with a missed reading rate of over 30%. Third, monitoring stations are installed without consideration for the diverse land use situations of each locality. Specifically, there are few stations along main roads where air pollution is heavy, but comparatively many in residential districts. For example, of 20 stations in Seoul, 13 are in residential districts, three are in semi-industrial zones, three are in greenbelts, and one is in a commercial zone. Pollution concentration readings therefore diverge considerably from the actual situation in each geographical area.

• **Environmental Impact Assessment System**

Provisions of the 1977 Environmental Conservation Law led to creation of the EIA system in 1982, and from that year to 1992 assessments were performed on 996 projects. The range of projects covered by EIAs at first included only urban development, industrial siting/lot grading, and energy development, but the range was gradually broadened so that it now covers 15 types including water resource development, port construction, railway construction, and waste management facilities. Since 1987 EIAs cover not only public projects, but those of private-sector companies as well. Especially numerous among the projects subjected to EIAs are energy development (power plants), housing developments, tourism development, industrial siting, and lot grading for industrial parks. But EIAs left a great deal to be desired, and a more full-fledged system did not come into being until after the 1993 Environmental Impact Assessment Law, which is based on the 1990 Basic Environmental Policy Law. According to researcher Choi Yongil, the EIA system until 1993 had the following problems: (1) It was impossible to cancel or modify project plans because EIAs were performed on the assumption that projects would be implemented, so assessments ended in after-the-fact rationalization of project plans; (2) basic data for environmental studies are insufficient; (3) agencies that approve EIAs have no interest whatsoever in implementing decisions made in discussions on assessments, so it is often the case they are not implemented; (4) environmental assessment criteria are controls on concentration, not on total pollutant load; and (5) EIAs do not sufficiently incorporate local citizens' opinions. That is to say, South Korea's EIA system has had problems similar to those of other Asian countries, including Japan. While the present system based on the 1993 law makes substantial improvements in these respects, Choi Yongil says that at the implementation stage procedures are still formalities. It remains to be seen if the new system will be effectively implemented

• **Emission Levy System**

South Korea's emission levy system, which was introduced in 1983, appears at first glance to incorporate the polluter pays principle, but this is actually not the case. Researcher Kim Infan points out the following problems. First, no emission levies are imposed for emissions of pollutants as long as they are below the concentration stipulated by standards. Emission levies are set by considering mainly the level of usable technology and the economic burden on businesses. Second, the criteria for calculating emission levies include (1) per-day pollutant emissions in excess of allowable levels, (2) duration of emissions, (3) levy coefficient according to the rate at which the allowable level has been exceeded, (4) levy coefficient according to number of offenses, and (5) basic levy amount, making it something akin to a penalty-like surcharge, and therefore ineffective as an economic incentive for reducing pollutant emissions. Third, pollutant concentration measurements are based on instantaneous readings taken during inspections, so levies are not proportional to actual pollutant emission amounts. And fourth, there are many problems in the way the system is run. South Korea has serious environment-related budget and personnel shortages, so that during 1992 each pollutant-emitting place of business received an average of only 1.8 visits for guidance and inspections. Furthermore, these visits do not include inspections of pollutant-emitting facilities and pollution-control equipment to evaluate their compliance, instead depending on the collection and analysis of pollutants emitted mainly via pollution-control equipment. Some have observed that, under these circumstances, it is more advantageous to pollute and pay a surcharge than to abide by standards.[5] In light of these observations, this emission levy system certainly appears inadequate as a practical arm of environmental policy. Since the 1990s South Korea has implemented a number of economic policy measures, such as an environmental improvement fee system, a waste deposit and fee system, a waste volume-rate system, and a water quality improvement fee system, but these too have a variety of problems.

---

[5] In fact, in interviews conducted by one of the authors (Cheong Deoksu) in the Ulsan area, a factory manager said he gets by with paying light fines even if he does not pay emission surcharges. Apparently there are some recent instances in which factories received criminal penalties.

4. The Promise of a New Political Milieu

4-1 Democratization and International Influence

With the arrival of the 1990s South Korea passed the Basic Environmental Policy Law (August 1990), and has achieved considerable progress especially in the area of environmental law. Of indisputable significance are changes in the political situation that underlie this progress: democratization under the Ro Taewu government beginning in 1987, the 1991 local assembly elections, and 1995 elections for local heads of government in conjunction with new local assembly elections. Especially characteristic is the rapid appearance of environmental NGOs along with the progress in democratization since 1987.

In the 1990s there has been a groundswell in pro-environment international public opinion, which was occasioned by the 1992 Rio Earth Summit. An important point is that many environment-related conferences, new advances in international interchange on the NGO level through these conferences, and other developments have considerably influenced the subsequent development of South Korea's environmental policy. For example, at the proposal of the Japan Environmental Council in December 1991, the first Asia-Pacific NGO Environmental Conference (APNEC) was held in Bangkok, where about 100 NGO representatives from the eight countries of Japan, South Korea, Thailand, Malaysia, Singapore, the Philippines, Sri Lanka, and Nepal met for the first time under the same roof, and the second conference was held in March 1993 in Seoul, where about 300 delegates from 10 countries met. During these same years international and interpersonal exchanges on the government level began among Asian countries, as with APEC and ASEAN environmental minister meetings, and the Environment Congress for Asia and the Pacific (Eco Asia) meetings held every year since 1992 under the auspices of Japan's Environment Agency. Through such interchange Asian countries are influencing each other considerably, and through this process South Korea's environmental policy has assumed a new orientation amid an increasingly international perspective.

4-2 The Rise of the Environmental Movement

The Korean Research Institute on Pollution Problems, founded in 1982 as what is arguably South Korea's first private environmental organization (an internal group of a Protestant church organization, its core members are church intellectuals, campaigners, and the like; it is now called the Korean Church Environmental Research Institute), has played an important role in dealing with Onsan disease and other problems. In those days, however, citizens' movements working on specific environmental issues were all regarded as anti-government or anti-establishment, so citizen initiatives were hampered by this disadvantageous political climate. But during the process of political democratization beginning around 1987 a variety of environmental movements sprang up and started their respective campaigns. These movements grew out of the milieu created by the democratization movement that existed under the long-term military development dictatorship, and was also part of that movement. Following is a brief look at a few of the major organizations.

South Korea's largest organization is the Korean Federation for Environmental Movement (KFEM), whose forerunner, the Federation of Movements for Eliminating Pollution, was created in 1988 by the coalescence of two other organizations founded a few years earlier. As similar organizations formed around the country through the 1980s and into the early 1990s, eight of these national organizations merged in April 1993 to form KFEM, whose main purposes are: (1) make the environmental movement a practical part of everyday life, (2) take the lead in eliminating pollution while working from within industry, (3) propose realistic policy alternatives, (4) maintain a consistent stand against nuclear weapons, and (5) strengthen solidarity with overseas environmental organizations, and look for ways to cope with problems internationally. As of 1996 KFEM had branches in 14 major cities around the country, including Seoul, Pusan, Ulsan, Taegu, and Kwangju, but these are at the same time all independent organizations. Nationally KFEM has well over 10,000 members, and the organization has its own office with about a dozen full-time staff members. KFEM is very active on a broad variety of efforts. Active as KFEM representatives and officers are many former anti-establishment intellectuals, academics, lawyers, clergy, and the like.

Another powerful organization is the Citizens Coalition for Economic Justice (CCEJ), which was founded in 1989 against the backdrop of citizen dissatisfaction and indignation over economic injustices, as symbolized in those bubble economy days by land speculation and improper accumulation of wealth. In that sense it was not an organization specializing in environmental problems. However, motivated by the impending 1992 Earth summit, CCEJ in 1991 established the "Center for Environment and Development" and started initiatives on development and environmental issues. CCEJ likewise has well over 10,000 members nationwide, 20 branches around the country, and its own office with a large full-time staff. It differs somewhat from KFEM in its emphasis on making policy proposals to government. Like KFEM, membership includes many academics, lawyers, and other specialists.

In addition to these two large groups, the South Korean YMCA and other long-established organizations now believe it is important to work on environmental problems. Many more organizations have appeared in the years since the late 1980s, such as the Hansamurin Association, the Korean Research Institute on Environmental and Social Policy, the Environment and Pollution Research Organization, whose primary members are researchers affiliated with Seoul National University, and Green Korea, a new environmental NGO whose office is in Taejon City.

5. Summation

The foregoing has constituted an outline and brief discussion of environmental problems in South Korea, and of the development, characteristics, and shortcomings of environmental policy for dealing with those problems. In general, except for a number of important differences, the situation in South Korea is in many ways like that in Japan, only running a few years later. In that sense, a 1995 analysis by Japanese researchers Harashima Yohei and Morita Tsuneyuki makes some interesting observations. They point out that there has been a 20-year time difference between Japan and South Korea in the process of economic development, but that with the process of environmental policy development that lag has diminished to between 12 and 14 years. Assuming they are right, South Korea has developed its environmental policy at a faster rate than Japan did at the same stage, which is perhaps the result of the "late-starter benefit" (see **Essay 2** at the end of this chapter). But if we consider the actual development of environmental policy to date, and if we assume that South Korea will be able to take advantage of this late-starter benefit in policy development, that challenge began only recently in the 1990s when the political situation changed. Furthermore, this will be possible only if South Korea learns from the successes and failures of the postwar Japanese experience and overcomes their limitations. In that sense, it is more important than ever for these two neighboring countries to share their experiences and conduct joint research pertaining to environmental problems and research.

(CHEONG Deoksu, TERANISHI Shun'ichi)

///

Essay 1   South Korea's Fee-Based Trash Collection Policy

Owing to economic growth and the rising consumption level since the latter half of the 1980s, the amount of waste generated in South Korea has increased yearly. For that reason the country now faces a critical situation because it has been disposing of 90% of its wastes by landfilling them without incineration, and now landfill space is running out. To deal with this situation the government at the end of 1993 announced its "National Comprehensive Waste Management Plan," which set forth three policy objectives for managing wastes, in the following order of priority: (1) reduction, (2) recycling, and (3) appropriate disposal. With a base year of 1992, an interim target year of 1997, which is the last year of the New Economic Plan term, and final objective target year of 2001, the plan's specific goals include: (1) reduce the per capita waste generated per day to 1.05 kg by 2001 and (2) increase the rate at which citizens pay for the disposal of their own wastes, thereby raising the rate at which local governments finance their own waste disposal, from 12% in 1992 to 60% in 1997 and to 100% in 2001. South Korea's policy of fee-based waste disposal was put into action with "reduction" as the key to success.

This paid collection policy was implemented on a test basis in 33 cities, counties, and wards throughout the country in April 1994, and then nationally on January 1, 1995. It is not a weight-based, but rather a volume-based system in which people are charged by the bulk of their trash put out in bags designated by law or ordinance, with fees obtained through the bag purchase price. The system serves residences and small businesses generating under 300 kg of wastes per day, and covers municipal wastes excluding briquettes, recyclables, and bulky wastes. At first the desig-nated bags were translucent plastic in 10-, 20-, 50-, and 100-liter sizes, but because the bags broke easily, 5-, 30-, and 70-liter sizes were subsequently added. Although bag prices vary according to the local government, at the outset in 1995 per-household monthly amounts charged by local governments averaged in the range of 2,000 to 3,000 won, which apparently differed little from pre-implementation fees, so from the latter half of 1995 to early 1996 all local governments have hiked their fees.

In 1996 the Environment Ministry analyzed and assessed the results in 15 cities and provinces and found that wastes generated decreased 26.7% (from 49,191 to 36,052 tons per day), while recycling increased 34.8% (from 8,927 to 12,039 tons per day). However, it also found that although household wastes decreased, business wastes continued to increase, and that with the passage of time after program implementation, the amount of wastes started to grow again. It is therefore too early to judge the effectiveness of this policy.

Meanwhile, local governments' income from fee-based collection is higher than before, which raised the fiscal self-support rate from 14% in 1994 to 30% in 1995. In FY1995 1 trillion won were budgeted nationwide for trash collection, transport, and disposal, of which 300 billion won were raised through the sale of trash bags. In this respect, South Korea's fee-based trash collection policy has definitely contributed to increasing the rate at which local governments finance their own waste disposal. This has effected a shift in the payment of waste management costs from general revenues to collection fees.

(SEONG Won Cheol)

Essay 2 "Late-Starter Advantage" in Environmental Problems and Policy

In economics we talk about the "late-starter advantage," which means, for instance, that countries getting a later start on industrialization can bring in technology, funding, and business management strategies from countries that industrialized earlier, thus benefiting themselves by making it possible to accelerate the process of economic development. But the downside of this is the "late-starter disadvantage."

What kind of significance does the late-starter advantage or disadvantage have in the areas of environmental problems and policy? Assuming there is a certain significance, what specifically is it? One report of great interest for examining five East Asian economies (Japan, South Korea, Taiwan, Thailand, and Indonesia) from that perspective and making an international comparison with regard to their development of environmental policy is *Managing the Environment with Rapid Industrialization: Lessons from the East Asian Experience*, a study written by David O'Connor for the OECD Development Centre. This study points out that East Asian countries have enjoyed a certain advantage through rapid industrialization that started late, and the greatest advantage was the opportunity to learn from forerunners. For example, there was less uncertainty about the absolute and relative magnitudes of environmental risks, which

was useful in setting environmental priorities. It was also possible they could choose from a wide variety of low-cost technologies in order to make them conform even more closely to various environmental purposes, and they were able to quickly incorporate the results of technological innovation.

This is indeed so, but a look at the actual situations in South Korea and Taiwan, which have come up from behind Japan in the area of economic development, shows that while they have considered Japan a model forerunner in terms of mainly development and technology, they have not necessary learned fully from Japan's bitter experiences in terms of environmental problems and policy. And even when it comes to Japan's few success stories, South Korea and Taiwan found it difficult to use their late-starter advantage because of hindrances such as the differences in political and economic institutions (local self-government, for example) that made those Japanese success stories possible.

In sum, in the areas of environmental problems and policy, developing countries must learn much from the experiences of their forerunners, but efforts at solutions and advances are defined to a large degree by each country's internal circumstances.

(TERANISHI Shun'ichi)

Chapter 3
Kingdom of Thailand

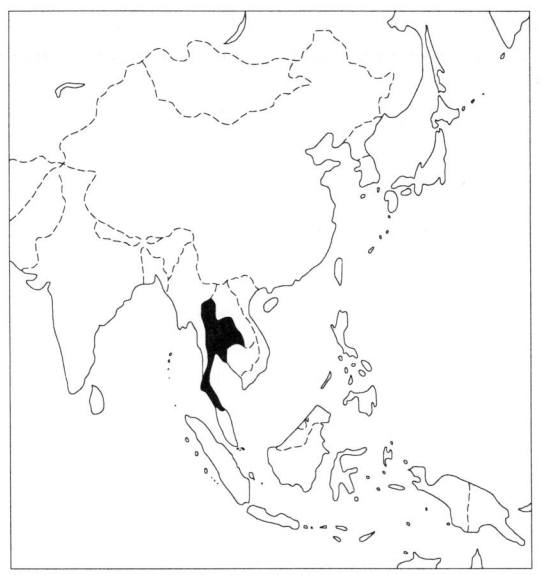

64

1. Introduction: Thailand's Economic Development and Environmental Problems

1-1 Plan Implementation and Rapid Urbanization

• Planned Development Using Foreign Capital

Thailand was on the vanguard of industrialization among Southeast Asian countries. Already in the 1950s the government had established an office to facilitate industrial activities and passed a law to encourage investment in industry, but it was from the late 1950s and into the early 1960s that it began full-fledged work on the infrastructure for industrialization and development. Specifically, in response to a World Bank recommendation the Thai government in 1959 created a system for development by establishing the National Economic Development Agency, then in 1960 passed the Act for Promoting Investment in New Industries and formulated its Regional Development Plan. The Economic Development Plan was formulated in 1960, and implemented in January 1961. Only then did industrialization begin in earnest. In 1964 the government set up its Rural Development Plan, which brought planned development even to the provinces. This Economic Development Plan served as the first step for development under the other above-named plans, but in 1966 Thailand initiated its first Five-Year National Economic and Social Development Plan. Although there have been minor delays owing to changes in government, the country is now on its eighth Five-Year Plan, which started in October 1996.

In accordance with these plans Thailand has built trunk roads and developed coastal areas. Despite the many changes in government, the Five-Year Plan has from early on been consistently formulated and followed in economic development. Among Southeast Asian countries Thailand deserves special recognition for sticking to this modern method. Another important element of Thailand's industrial development policy to date is the active use of foreign capital. The 1960 Act for Promoting Investment in New Industries allows 100% foreign-owned firms, and except for a brief hiatus government policy has consistently promoted development using foreign capital from the time this law was passed. Starting in the 1970s the Thai economy has advanced rapidly, but because much of that has been achieved using foreign capital, plan implementation has been greatly influenced by foreign capital trends. At the end of the 1980s Japanese investment accounted for about half of all foreign investment.

• Growing Urban-Rural Disparity

Although economic development was planned, there was a considerable divergence between plans and their implementation. For example, "correcting inter-regional disparities," "balanced development of rural and urban areas," and "correcting the disparity between cities and farming villages" have been a permanent part of plan agendas. Especially since the fifth Five-Year Plan the "balanced development of rural and urban areas" has been a goal, but the disparity in fact contin-

ued to grow. In 1960 when planned development started, 82% of working people were farmers, which dropped to 73% in 1980. Meanwhile, the agricultural portion of the GDP declined from 40% to 23% during those same two decades, and in 1990 it was 15%. This trend encouraged the population to concentrate in the cities, and in 1990 only 57% of the working population was engaged in farming. From the 1960s and into the early 1970s migration from the country to the cities was mainly within certain regions. For instance, Thailand's second-largest city Chiang Mai rapidly urbanized beginning in the 1970s to a population density of 2,500 per square km, but subsequently migration bypassed outlying cities and increased the number of people moving directly into Bangkok. Thus, over 10% of Thailand's total population came to live in Bangkok, in which the population and economy increasingly concentrated. Bangkok also attracts far more than the other cities, which is because investment is overwhelmingly concentrated in the Bangkok area, where over half the country's factories are located. Thailand's northeast region has always been poor, and in the 1980s its per capita GDP was still 40% that of the national average. Other than the greater Bangkok area and its environs, the per capita GDP is dropping yearly in other regions as well. In all countries population tends to concentrate in the cities, but it is especially pronounced in Thailand.

• Lack of Overall Balance

One factor bringing about this situation is development "project-ism." There is no doubt that projects have elicited economic vitality, brought about development in a way that surpasses the predictions of the five-year plans, and maintained a high economic growth rate, but that has not necessarily been coupled to a comprehensive development plan. Instead, various giant projects have been carried out independently of one another. Not a few of the public works projects for infrastructure development have also been funded by foreign capital, which has led to the selection of profit-making projects such as electricity generating plants and expressways, while non-profit-making plans languish. Investment like this has had a negative environmental impact. Particularly since the 1980s, foreign currency-based investment shifted from the public to the private sector, so there has been little of the investment in railroads and other things needed to hold down motor vehicle pollution. Reckless development occurring mainly in big cities, and the construction of roads and railways in Bangkok as a means of resolving traffic congestion, are likewise conducted project by project with primary consideration for what the implementing agency has to gain. For this reason urban environments deteriorate, and Bangkok's traffic congestion and motor vehicle pollution are not alleviated. This is not simply due to inappropriate management of plan implementation because in addition the inadequate regulation of land use makes it impossible to put the brakes on economic development, which is another major factor that hampers the proper disposition of projects and prevents the balanced development of localities.

1-2 Assisting Rural Development

• The Downside of Agro-Industry

Notwithstanding the gradual decline in agriculture's share of GDP, agricultural products were Thailand's primary exports until the 1990s when their first-place ranking was supplanted by industrial products. Agriculture for export had its origin in the agro-industry proposed by the fourth Five-Year Plan, which started in 1970. Low rice productivity has been considered a problem of Thai agriculture, and even now it is not very high. The government therefore encouraged agro-industry as a component of its agriculture promotion policy, which led to the cultivation of rubber, sugar cane, and other monocultures as tropical crops to replace the production of grains, mainly rice. Developing sugar cane, cassava, and even broilers into export industries made agro-industry into the most important industry for domestic capital. In the 1980s Thailand started shrimp aquaculture and rapidly increased its exports to Japan. This agro-industry example is one of the few success stories for the Five-Year National Economic and Social Development Plan. In Thailand's economy the growth of these sectors has also been a factor summoning forth domestic demand and raising the GDP, but there was a downside to this agro-industry growth: As early as 1973 the effluent from a sugar factory near Kanchanaburi seriously damaged fisheries and caused other problems such as deforestation and water pollution.

• Aiming for the Indochina Market

Rectifying the disparity between urban and rural areas has always been a main item on the planning agenda, and all five-year plans have emphasized development especially in the poor area of northern Thailand in an attempt to eliminate that disparity. Beginning in the second half of the 1980s the government reinforced its policy of dispersing industrial parks through the outlying districts, but investment is still centered mainly around the capital, which has made poor areas even more so and pressed northeast Thailand even further toward development-intensive thinking. Especially in recent years both the central government and outlying regions have been planning many industrial parks for northeast Thailand in the belief that it can become an industrial base aimed at the entire Indochina market. Under this policy of priority to development, even foreign capital is moving into the provinces in search of cheaper labor, and already a variety of problems has arisen (see Essay 2 at chapter end).

While development in northern Thailand is oriented toward Indochina, this also includes development in the three-country region of northern Thailand, Laos, and China's southwest region. Northern Thailand has already seen a great deal of dam development. Electric power development in particular is the main element of development plans in this region, but throughout Thailand — including this region — dam construction is gradually becoming more difficult because environmental and social problems are coming to the fore. Although hydropower does not account for a large proportion of electric power, a stable supply of electricity is

considered vital to Thailand's economic development, and so building generating capacity in other countries is seen as one solution. Thailand buys electricity from dams in Laos, and a central part of this plan is further power development in Laos and southwest China. Contracts have already been signed for transmission of 1,500 MW of power from Laos by 2000. Formerly it was a case of foreign capital investments in Thailand causing environmental problems, but in recent years Thai capital has begun adversely affecting the environments of neighboring countries. Thai capital has already made many investments in China, where the problems it causes can no longer be overlooked.

• Development Plans and the Environment

As its name shows, the Five-Year National Economic and Social Development Plan is not concerned with economic development alone, but in fact its main purpose is increasing the GDP. Still, this is not to say that the plans are totally unconcerned with preventing environmental degradation, i.e., the negative aspects of development. The fourth plan incorporates environmental conservation. Further, water pollution by sugar cane processing factories, pollution caused by Japanese-owned caustic soda plants, and other problems occasioned the addition of provisions for nature conservation and pollution prevention to the constitution in 1974 and the creation of the National Law on Preserving and Improving Environmental Quality, under which the National Environmental Board was established. In 1981 the Cabinet approved Board guidelines called "Policy and Measures on Development Heedful of the Environment." But these plans and measures have hardly been effective in the face of large-scale development projects.

The sixth Five-Year Plan (1986-1991) was the first to earnestly incorporate and put efforts into environmental measures. The seventh plan (1991-1996) set forth clearly defined goals for environmental quality in order to raise the quality of the environment and natural resources. The year 1992 in particular was an important juncture in Thai environmental legislation. Under the seventh Five-Year Plan the National Environment Act was completely overhauled, laws including the Factories Act and Hazardous Substances Act were formulated, and environmental policy was fortified (see **Essay 1** at chapter end).

2. Environmental Costs of Rapid Urbanization and Industrialization

2-1 Urbanization and the Environment

• Motor Vehicle Pollution

The concentration of the population and businesses in the cities, especially in Bangkok, is bringing about grave motor vehicle air pollution. Bangkok's traffic jams and vehicle pollution are now world famous. The city's concentration of people and economic development have rapidly increased its number of vehicles. While in 1979 there were

6.1 million registered vehicles, this exceeded 27 million in 1993 owing to rapid economic development lasting through the 1980s. But 42% of these vehicles are two-wheeled, and the majority of those have two-stroke engines. Road-building cannot keep up with the swift increase in vehicles, so the provision of roads and traffic signals, and the enacting of one-way traffic restrictions, have proceeded on a catch-up basis. Exhaust controls are behind, and especially because vehicles in use have hardly been maintained, they pollute badly. The large number of two-wheeled vehicles also worsens air quality. Particulate and hydrocarbon concentrations are particularly high, and lead pollution is pronounced because leaded gasoline has been used. Examples of airborne particulate pollution are 0.86 mg/m^3 in Yawara and 1.3 mg/m^3 in Pat nam , and readings at 14 of the 15 roadside monitoring stations exceed the standard of 0.33 mg/m^3. The large number of vehicles contributes greatly to particulates, but the urban construction rush also produces a sizable amount. Lead contamination is 0.68 mg/m^3 in the Shipuraya and Opharai areas. Some confirmation of the gravity of air pollution comes from measurements, but the only source of information on health damage is a health survey of traffic police officers. While estimates have been made to an extent, the government has not conducted studies to determine citizen health damage. Such studies require funds and trained personnel, and the situation demanded that studies begin with training the personnel. With help from the World Bank, plans for a study of health damage to children finally got off the ground in 1996. To deal with motor vehicle exhaust the government has been establishing by degrees a program to test emissions in order to require maintenance of vehicles in use, and it has begun taking steps to prohibit leaded gasoline.

• Municipal Wastes

Concentration of people in the cities highlighted the waste problem. Moreover, rising GDP changes people's consumption patterns, which increases the generation of household wastes faster even than population growth. In the principal cities of Bangkok, Chiang Mai, and Phuket, household wastes are growing faster than any other kind. Although cities collect household wastes, their capacity to dispose of them cannot keep up with the volume. Bangkok collects 4,530 tons of wastes daily, but it has the capacity to dispose of only about 10%, which underscores the urgency of building waste management facilities. Additionally, the large amount of plastic, which is characteristic of modern wastes, challenges municipalities to develop new systems for waste treatment and recycling.

2-2 Industrialization and the Environment

• Air Pollution by Industrialization

Industrialization has swiftly advanced, especially since the 1980s, and it was the construction of many industrial parks that provided the infrastructure. Since 1972 Thailand has created over 50 industrial parks. Foreign capital went into mostly the export processing divisions of these parks, and that served as the driving force for economic growth. But progress in industrialization far outpaced the building of a framework to prevent air pollution, resulting in pollution that worsened steadily as industrialization marched forward. Current law establishes environmental standards for the atmosphere, and there are emission standards for each type of fixed source covering SO$_2$ and other pollutants. Penalties are levied for violating emission standards, and under the Factories Act factories can be shut down and issued orders to make improvements.

In South Korea and Taiwan air pollution by the heavy and chemical industries is severe, but in present-day Thailand, by contrast, pollution by small and medium-sized factories is particularly bad. Many of these facilities are not regulated, and in many cases they operate without any pollution control equipment at all. Thus, even if one can easily imagine their heavy atmospheric pollutant emissions, there are no data from measurements of those emissions. Data are not available even for some large facilities.

Furthermore, in the Bangkok area and densely congregated factory zones on its outskirts, or in outlying industrial cities, there is a combination of factory and motor vehicle pollution. Samut Prakan near Bangkok, for example, is one of the largest factory areas of the country and has particularly poor air quality. What is more, the castor oil manufacturing plant in Samut Prakan Province emits dust and foul odors that adversely affected the respiratory systems of people living within 2 km of the plant, which happened because the plant has no pollution control hardware.

Currently power plants are the fixed sources emitting the largest amounts of the air pollutants NOx, SOx, and CO, and for this reason air pollution has caused health damage around the power plant in Lampang Province. In September 1992 pollutants emitted by a lignite-fired power plant caused the hospitalization of three-fourths of the surrounding residents with respiratory ailments, and created 1,000 outpatients. There were also domestic animal deaths and crop damage. The power plant's measurements showed an atmospheric SO$_2$ concentration of 2,100 mg/m^3. Although the plant did have a dust collector, its capacity was exceeded by the amount of coal burned, and the collector was not designed to remove SOx and NOx. This generating plant had no pollution control equipment despite its being state-run, a fact that candidly illustrates the current situation with air pollution control in Thailand. Natural gas fires 44.09%, oil 27.95%, and lignite 21.30% of thermal power plants, ranking lignite third, so these lignite-fired plants, which account for over 20% of the whole, constantly present this danger. After this Lampang incident, however, the government required all new power plants to install equipment to remove SOx and other pollutants. A matter of note is that victims and the power plant reached a settlement in which pollution damage is certified and compensation paid.

• Hazardous Wastes

Since the 1980s hazardous wastes have also increased quickly. Their rate of increase is higher than the economic

Table 1 Thailand's Forested Area (1961-1993)

Figures are millions of *rai* (total area of Thailand is 320.7 million *rai*)

Year	Forested area	Decrease in area	Percent of country forested
1961	171.0		53
1973	138.6	18.95	43
1976	124.0	10.53	38
1978	109.5	11.69	34
1979	97.5	10.96	30
1985	94.3	3.28	29
1989	89.6	4.98	27.95
1991	84.7	5.50	26.60
1993	83.5	1.42	26.04

Source: Thailand Forestry Department.

growth rate, with the amount generated in 1995 estimated at 22 million tons. Many hazardous wastes come from factories and other business establishments, and account for about 80% of wastes generated. Most come from establishments such as metal processors, electric appliance manufacturers, and chemical manufacturers. By region, the overwhelmingly large proportion of wastes comes from the 10 provinces around Bangkok, which host over 50% of Thailand's factories. Often these hazardous wastes are given slipshod management at best, or just left anywhere or dumped illegally. To avoid such situations and deal with hazardous waste management, the Industrial Waste Treatment and Disposal Center was built in Bangkok in 1988. This facility, the first of its kind, is presently the only operating hazardous joint waste management center in which the government is involved. The facility handles solid wastes, sludge, and waste liquids, and it receives wastes from industries such as metal processors and tanneries, which produce particularly large amounts of hazardous wastes. Treated wastes are taken to a final disposal site, which is in a different location. Four facilities were planned, but the incinerating and chemical treatment facilities have been either delayed or canceled. Although nearly completed, the chemical treatment facility has met with stiff citizen opposition that has stalled the project. Citizens are very concerned about whether industrial wastes will be treated in a proper manner.

Other industrial wastes are likewise on the increase, but there is no public involvement in such wastes, and they are managed by those who generate them. In 1992 the Factories Act required businesses to manage their own industrial wastes, but owing to the delayed construction of final disposal sites and other factors, proper management of industrial wastes as a whole is a serious matter, not to mention the inappropriate management of hazardous wastes.

3. The Environment in Rural Areas

The loss of forests is representative of the development and environmental problems in rural areas where primary industries predominate. Development in these regions exacts a price that is manifested as deforestation.

In 1960 when the economic development plan was implemented Thailand's forested area was 27,362.9 ha, corresponding to 53% of the country. Since that time, however, deforestation has proceeded with unusual swiftness as economic development advanced. Now about half of the forests existing in 1960 have vanished, shrinking to about 26% of the country's area (Table 1). In the south, which originally had little forest cover, forested area is down to about 10%. Only 834,500 *rai*[1] of old growth forests remain in 79 reserves. This rapid change has had a great impact on ecosystems. Additionally, the loss of aquifer recharge forests has brought about severe shortages of water for drinking and irrigation, and is also closely linked to global warming.

3-1 Forests Converted to Farmland

There has been much debate on the reasons for this deforestation. Just as in other Southeast Asian countries, the wood industry is one of Thailand's main industries, and so it naturally accounts for much of the loss, but the proportion it accounts for is not that high. It is the agro-industry mentioned above that has deforested much land. Agro-industry has not only cut down inland forests, but also many coastal mangroves in order to farm shrimp. Additionally, it has indirectly deforested land through its purchases of much farmland because farmers chased off their land by these purchases have been forced to go into forested areas and cut trees in order to secure new farmland. Further, in the northeast and other regions where productivity has always been low, farmers cut trees to enlarge farmland and increase their incomes. At present nearly all the forests that are convertible to farmland have been cut, and it is no longer possible to increase income by expanding farmland. As over 70% of the area deforested since the 1960s has been converted to farmland, conversion has been a central and continuing cause of deforestation related to the primary industries.

---

[1]   One *rai* is 1.12 ha.

3-2 Deforestation by Construction and Tourism

Although the primary industries account for the major share of deforestation, other causes also bear examination. One example is the building of infrastructure like roads and dams. Public works like these are an important element of the Economic and Social Development Plans, and have precedence over protecting forests. And as one can see from Thailand's dependence on tourism for 17% of its foreign currency earnings, its tourism industry is also growing rapidly. There are increases not only in the foreign tourists who bring foreign currency, but also in Thai tourists, whose number grows with the economy. This growth in the tourism industry propels not only the development of existing urban areas, but also of inland and coastal areas, and thus deforestation by tourism development is a serious matter. Forests are cleared for hotel construction and for building the roads that facilitate the tourism industry, but added to this since the 1980s is the fast-paced construction of golf courses. Although a small part of the total, they are an undeniable factor in deforestation during recent years.

3-3 Government's Response

In order to put the brakes on deforestation, the Thai government in 1989 banned logging in general, and established a concession system for commercial forest development, under which only licensed people may log, and only within their concessions. But in 1988, before the logging ban, the government granted 317 concessions covering 18.13 million ha. What is more, many of the licensees were government-affiliated forestry organizations. When implementing the seventh Five-Year Plan the government emphasized the protection of forests and afforestation. So while on the one hand allowing logging under concessions in commercial forests, the government at the same time required afforestation under the Afforestation Law, and it also developed a National Afforestation Plan for the afforestation of logged-over areas and for encouraging tree-planting by local citizens. Other measures include the building of barricades to keep people out of the forests. And the farmers themselves, whose attention was wholly occupied with the expansion of farmland by cutting trees, have begun to see that cutting the forests is causing environmental problems, so that now afforestation is conducted at the local level with NGO cooperation.

The creation of the logging concession system and the implementation of forest protection policies brought about a quick drop in the deforestation rate, and immediately after the general logging ban took effect the deforestation rate was a mere 0.3%. This does not mean, however, that logging has stopped; the loss of forests still continues, and afforestation activities cannot keep abreast. Even since the general logging ban, Thailand's area of forested land continues to decline at an average annual rate of 1.6% owing to the factors described above. The main reason for this is that the inadequate monitoring system fails to spot continued illegal logging. In addition to loggers cutting trees illegally outside of their licensed areas, another problem is that people use the logging roads in concessions to move farther into the hinterlands, where they log illegally.

3-4 Vicious Circle in Northeast Thailand

While northeast Thailand is one of the country's most forested areas, it is also its poorest, with low agricultural productivity. Thus in recent years efforts have made to diminish the widening income differential by increasing the area of arable land, resulting in continued illegal logging. In 1953 northeast Thailand had a forested area of 641,000 *rai*, accounting for 60.5% of the whole country's forests. But two decades later in 1973 it had lost 342,000 *rai*, leaving a forested area of 297,000 *rai*, which declined further to a mere 136,200 *rai* in 1991. This region has a much faster deforestation rate than others. A direct reason for illegal logging by farmers is the inadequate monitoring system, but a more profound factor is poverty and the maldistribution of wealth. In order to stop this deforestation Thailand's government has therefore, in addition to its policy for encouraging agro-industry, pursued a policy toward agriculture that gets farmers to increase yield per unit area, and move them away from increasing income by expanding farmland area. While this policy has achieved a measure of success, farmers are still caught in the vicious circle of trying to increase their farmland because of the continually widening income differential with urban areas.

4. Rivers

Presently water is the most urgent issue throughout Thailand. Water presents a challenge that is common to all developing countries owing to the environmental problems caused by securing water supplies and to the pollution of major rivers, and in Thailand it has become particularly serious as concentration into the cities and industrialization proceed rapidly.

4-1 Securing Water Supplies

Thailand's national average annual rainfall is about 800 trillion cubic meters, an amount that belies the possibility of water shortages, but in fact the country has suffered from them every year since 1987. This is because of the rainy and dry seasons, a peculiarity of Thailand's climate. During the rainy season much rain falls, but in the dry season water shortages have become especially acute. In 1993 the problem came to the point where the government had to mount a water conservation campaign to resolve the shortage. Although the campaign achieved some success, the truth of the matter is that the heavy rains in 1994 allowed Thailand to avoid the worst.

Water demand breakdown is 20 billion cubic meters for household use and 10 billion for industrial use, amounts that contrast with about 4 trillion for agriculture. From these figures one can see the large proportion for agriculture, and indeed, the eight National Economic and Social Development Plans have all placed importance on securing agricultural water supplies. Especially the fifth Five-Year Plan included development plans concerning primarily agriculture, and securing irrigation water was a particularly big item on the agenda. At the same time, however, advancing industrialization called for a stable industrial water supply, and rapid urbanization has necessitated finding water for the cities as well. Especially the golf course construction rush since the 1980s has created heavy water demand that now competes with agricultural demand. Thailand's steadily growing water demand is such that satisfying demand or alleviating shortages is a matter of national concern. Although the water shortage itself is not an environmental problem, it is both the cause and the result of environmental damage. For instance, one cause of the chronic water shortage is excessive logging in catchment basins. Loss of forest cover in aquifer recharge areas has reached crisis proportions.

Dams and reservoirs are being built to satisfy water demand in a bid to cope with this water shortage. Even the existing 25 large dams are insufficient to solve the water shortage, so many more dams are on the drawing board. Despite the fact that logging of forests is one cause of the shortage, a vicious circle is in the making because forests are cut to build more dams. Such dam construction not only harms forests and wildlife, it has also forced residents of dam construction areas to relocate, and destroyed their means of livelihood. There have been instances in which people forced to give up their land by dam construction mounted opposition campaigns. As it is doubtful that the harm caused by dam construction is offset by alleviation of water shortages or the supply of electricity, there is controversy over whether building large dams is a solution in the true sense of the word. Of particular note is that in some areas there are new efforts at reducing dependence on large dams for irrigation water, estimated to be the largest category. It involves using NGO help to combine the building of reservoirs in each area with traditional methods of irrigation and water use. With hydropower, however, the trend in recent years is responding to electricity demand by building dams in Laos and China.

Another cause of water shortages is free tap water. The citizens' mindset is therefore that water is free and will never

run out. The same goes for factories, which have many water-intensive production processes and no incentive to change over to water-conserving processes. This total lack of demand-side management is certainly a major part of the problem.

4-2 Pollution Extent of Major Rivers

Some of Thailand's principal rivers are the Chao Phraya, Tachin, Mae Klong, and Panpakhon. Each river's catchment basin has different characteristics and each river's state of pollution differs (Table 2).

• Chao Phraya River

We should begin with the Chao Phraya River, which flows through the Bangkok area and other areas with high population density and intensive land use. The river itself is also intensively used through its entire watershed. Its water is used for agriculture, drinking water, and industry, and the river also hosts much fishing and boat traffic. Four principle rivers in northern Thailand, the Ping, Yang, Yom, and Nan, converge in Nakhon Sawan Province where they become the Chao Phraya. This is Thailand's largest river, with a length of 980 km and a watershed of 177,000 square km. It runs through central provinces, Bangkok, and then into Samut Prakan Province. The river is composed of upper, central, and lower portions.

Thailand's National Environmental Board has been measuring the river's biological oxygen demand (BOD) since 1980, and since 1983 taking cadmium, lead, and mercury measurements. Results show that the Chao Phraya is heavily polluted, although that is mainly in the central and lower portions, while the upper portion, which runs through a farming and fishing region, has negligible pollution in a few locations. Measurements in 1993 showed that water quality was altogether good over the 312 km from Ampho Muang Nakom Sawan to Ampho Muang Nonthaburi. By BOD the amount of effluent entering the upstream watershed is about 10 t/day. In the central portion the river starts getting polluted at a considerable number of places. Water quality worsens drastically in the area from Ampho Muang Pathum Thani to Nonthaburi Province, because this is where the industrial district begins. Effluent increases to BOD 22 t/day. Household effluent is considered the main source of pollution in the central portion. Water quality in the lower reaches worsens substantially due to factory effluent. In the 55 km stretch from

Table 2 Water Quality of Major Rivers in Thailand

River	Standard	DO (mg/l)	Standard	BOD (mg/l)	Standard
1. Upper Chao Phraya	2	6.3	<6.0	0.8	>1.5
2. Central Chao Phraya	3	5.0	<4.0	1.4	>2.0
3. Lower Chao Phraya	4	2.2	<2.0	2.4	>4.0
4. Mae Klong	3	5.8	<4.0	0.8	>2.0
5. Nam Phong	–	4.6		2.0	
6. Tapi-Phum Duang	–	5.1		2.0	

Source: Department of Pollution Control, 1994-1995, Thailand.

Nonthaburi Province to the river's mouth at Ampho Phra Pradaeng in Samut Prakan Province, dissolved oxygen (DO) is under 2 ppm. Effluent in this region is BOD 217 t/day. In some places the water is so polluted that DO is zero, meaning that water quality worsens seriously as it flows through the Bangkok area, which has high concentrations of factories and people, and it is especially bad in the dry season.

Factory effluent contains chemical substances and heavy metals, the latter being more or less within environmental standards. But heavy metals have found their way into drinking water on multiple occasions. Average levels of the organochlorines used in agriculture also satisfy standards, but measurements of these hazardous substances are concentrations in bottom mud. Their accumulation and action in the food chain are of concern. Another problem is that cholera and other diseases afflict people who drink this water.

• **Mae Klong River**

The Mae Klong River is one of Thailand's major rivers in the country's west central region. Its 140 km length flows through Kanchanaburi, Ratchaburi, and Samut Songkhram provinces. The upper portion is the 95 km passing through Kanchanaburi and Ratchaburi provinces, and has Thailand's heaviest concentration of sugar cane plantations. This area's 17 sugar factories have over 60% of Thailand's total sugar production capacity. Sugar cane is a primary item in the country's agro-industry, which exports much sugar. These factories were built between 1969 and 1972, and as soon as they began operating, the river quickly became polluted, killing fish and other aquatic life because nearly all sugar factories lack effluent treatment facilities, and have directly run so much sewage with large amounts of organic matter into the river that natural aeration is impossible. To solve this problem the Ministry of Industry in July 1974 finally started an effluent treatment project, which involves requiring sugar cane processing plants to build effluent treatment facilities and government financial assistance to build the facilities, with that government money going to a fund used for 10 years' operation and maintenance expenses. At the end of the decade the factories repaid the building expenses and took over facility maintenance and control. Then two decades after the project in 1994, the Ministry of Science, Technology and Environment reported that river water quality had returned to normal.

The Mae Klong's lower portion begins in Ratchaburi Province, flows through Samut Songkhram Province, passes through a densely populated area, and empties into the sea. This area has many factories for fish sauce and other products. As these small factories have inadequate or no effluent treatment, water quality worsens still more. Currently the industries mainly responsible for poor water quality are sugar refining, and food processing such as fish sauce. Many of these factories are small and therefore not subject to regulation, but their large number demands effluent treatment.

• **Nam Phong River**

This river runs through northeastern Thailand. Most of its watershed lies in an agricultural region with cassava, jute, sugar cane, and other crops. Water to irrigate these plantations, as well as to supply water to people in the southern part of this region, comes from the Nam Phong. The Nam Phong is also used for bathing. There are two water treatment plants on the river that supply drinking water for the watershed's inhabitants, but people who live outside these plants' supply areas drink untreated river water. As northeastern Thailand is a poor district, the government promotes the siting of factories here in order to alleviate poverty, and the Ubonrat Dam was built on the Nam Phong River to serve as infrastructure for factory siting. Construction of the Friendship Highway network joining the region with Bangkok, railway construction, and the provision of other infrastructure has brought to the watershed in Khon Kaen Province sugar factories, paper factories, 174 small and two large rice mills, natural gas wells, and the like, plus many cottage industries such as rug factories. These factories constantly pollute the river.

While sewage from factories and homes is a continuing problem, serious pollution also comes from accidents in many instances, and the effects are manifested immediately in the dry season. Typical of this was a 1992 fire that broke out in a plywood factory that was using the sugar cane residue from sugar manufacture. During the firefighting, which took more than a day, water was sprayed on the factory's stock of bagasse for plywood manufacture, and that water drained into the river. This in turn forced from an effluent treatment pond the washwater used to clean a sugar refinery molasses tank. Additionally, molasses itself leaked from the joint of a damaged tank. With the Nam Phong's water level down in the dry season, the pollutants stagnated, and large numbers of the river's 38 fish species died and rotted. Five days after the accident this pollution entered the public water supply in Khon Kaen Province. Because this was of course unsuitable for drinking, on the sixth day the Ubonrat Dam released 10 times the ordinary flow for a week in order to flush away the pollutants, but this only worsened the damage. In the end the watershed covering 2,000 villages in six provinces 30 km downstream was polluted. During the 24- to 48-hour residence time of the polluted water, bacteria multiplied, the molasses broke down, and river water DO fell to zero. Polluted water flowed 70 km downstream, killing nearly all of over 60 fish species, which of course was a blow to the fishing industry. Contamination of irrigation water severely damaged crops, and public water plants were unable to take in water for over a day. In court the incident was judged an accident and the factory was not held liable for negligence, but it did pay damages and shut down for 180 days to make improvements.

This case shows that pollution released into rivers during the dry season can cause untold harm. As primary industries are severely affected, in this instance farmers retained lawyers and prepared to file lawsuits demanding redress. Other citizens also organized and conducted anti-pollution campaigns. NGOs became involved by performing damage surveys and informing citizens of their rights. Movements by farmers and other citizens were mounted precisely because of the role played by NGOs.

But as time passed the movement subsided, and the farmers gave up filing suit. One reason is that when the rainy season came the pollutants were all washed away, making it look at a glance as if the problem had been solved. Another critical factor was that many of the farmers grow sugar cane and to make a living must sell it to the sugar factory that caused the accident.

• Tapi-Phum Duang River

The Tapi River flows through Surat Thani Province in the south. On the lower portion the Sok and Sab canals join to become the Phumduang Canal, which merges with the river. The river flows through an agricultural district, but the area were it joins the canal is a factory zone. The Tapi had good water quality and was rich with aquatic life, but quality deteriorates year by year. Graywater and the effluents from fish markets and other markets are an especially great burden. Factory effluent, much of which is raw, enters the river at the confluence. On June 30, 1987 aquatic organisms in the Phumduang Canal died from oxygen depletion and floated to the surface, an incident that was repeated that August and in January and February of the following year. This pollution spelled disaster for fishing, one of the province's main industries. Furthermore, this river water was taken into the water treatment plant, forcing the plant to use more chemicals in treatment. A survey by the National Environmental Board and the Ministry of Fisheries narrowed the cause down to three possibilities. First was polluted water released from the Rachapurapa Dam; second was effluent from factories along the river, including a state-run brewery; and third was an influx of sea water. Local citizens believed it was the brewery because it had no wastewater treatment facilities and ran its wastewater directly into the river. Ultimately the problem was more or less solved by requiring the brewery to install effluent treatment facilities.

4-3 River Pollution: Main Problems and Future Challenges

A matter of important note is that Thailand's rainy and dry seasons make for differences with other countries. Specifically, heavy river flow in the rainy season will perhaps make water quality good, but in the dry season that becomes a dribble, and water quality deteriorates. It is often the case that average annual water quality is good, but is very poor at certain times, which calls for special attention to dry season data.

Heretofore there have been no reports in particular of health damage caused by river pollution, but as all major rivers are polluted, even rivers formerly considered unpolluted are now unfit as drinking water. Of course this means that aquatic organisms have an increasingly inferior habitat, but even greater are the concerns about damage to human health. Although no health damage is evident at this time,

the Thai Development Research Institute (TDRI) estimates that illness traced to Chao Phraya River water is costing the government 9.3 million baht in medical expenses. Whatever the case, the pollution of major rivers as described above imposes vital tasks upon Thailand in terms of the following three points.

First, large amounts of graywater from Bangkok and other densely populated areas enter the rivers untreated. As household effluent is BOD 212 t/day, untreated household effluent plays a major role in worsening river water quality. Dealing with this will not be simple.

Second is regulating factory effluent, which has caused heavy livelihood and asset losses. Because the Thai economy has developed mainly around the agricultural and marine products processing industries, effluents from these processing plants wreak a great deal of damage. Thus even without an accident, agro-industry effluent generates a heavy pollutant load. Outside of agro-industries such as food processing and brewing, the heavily polluting industries are papermaking, rubber, and tanning. Even as far as the government knows, about 30% of all factories have hardly any effluent treatment equipment. Many of these small and medium-sized factories are not regulated. Monitoring shows that, although in very small amounts, hazardous substances such as mercury, lead, and cadmium are released into rivers. Although at present there are no reports of direct damage from heavy metals, there are concerns about pollution from their accumulation.

Third is water pollution caused by agriculture. Major causes of pollution in Thailand's rivers are insecticides, chemical fertilizers, and the chemical feed in wastes produced by shrimp aquaculture, and by domesticated animals and fowl. Thailand also has many mines, but there is insufficient awareness of their pollution.

5. Summation

As this general survey has shown, Thailand's environment is rapidly deteriorating as development and the economy advance. Forests disappear as one watches, rivers are contaminated, and pollution incidents are frequent. What is more, there is still insufficient understanding of the extent and state of pollution. For example, Bangkok has only seven round-the-clock air pollution monitoring stations, and hardly any studies have been made of air pollution-induced health damage. Thus, hardly anything is known about the hidden damage already occurring.

Additionally, the state of deforestation and pollution reveals a lack of systems and personnel to deal with these problems. Despite the existence of legal emission standards or prohibitions, monitoring systems are inadequate. For the time being, a vital task will be to make up for this lack by providing citizens with information and making them capable of monitoring. For that purpose Thailand needs further democratization as well as greatly expanded citizen participation and local autonomy.

(Sunee MALLIKAMARL, ISONO Yayoi)

Essay 1   Thai System for Environmental Protection and Pollution Abatement

Thailand's National Environmental Board and Ministry of Science, Technology and Environment form the core of its organization for environment-related administration. The former is empowered to set environmental and emission standards for air, water, noise, and vibration, and to propose nature conservation plans to the Cabinet. It has the authority to formulate a broad range of environmental policy and is assigned the role of monitoring administrative actions on the environment. Its subordinate bodies, the Environmental Policy Planning Office, the Pollution Management Office, and the Environmental Quality Improvement Department, divide the Board's authority among themselves. The Ministry of Science, Technology and Environment implements environmental controls.

1992 was a milestone year for environmental law in Thailand. The 1972 National Law on Preserving and Improving Environmental Quality, a basic environmental law, was revised in 1992 to endow it with provisions to penalize violations of controls, and an environmental fund was established. Enacted in 1992 was the National Environmental Quality Act, which required EIAs and created a system for citizen participation. The Factories Act establishes controls on factory pollution. The Minister of Industry has the authority to determine standards and control restrictions in order to regulate environmentally harmful effects of factory operation such as wastes and pollution. Factory operators are required by the 1992 Factories Act to make environmental measurements, record the readings, and report them to the Ministry of Industry once each quarter. When factories commission waste management, the management companies are under the same obligations. Under this law, government pollution controllers have the right to make on-site inspections, and they can also take samples. When necessary they also have the authority to issue orders for improvement, and to impose penalties for infractions. In addition, they are empowered to offer advice to administrative officials who are authorized to order factories to shut down, and to suspend or cancel operating permits. Other related laws include the Hazardous Substances Act, Forest Law, Afforestation Law, and the Public Health Law.

EIA reports must be submitted when carrying out new, large-scale development projects that might affect the environment. These reports must specify methods of forestalling environmental impacts and the measurement methods used for actions that affect the environment. Reports are approved after examination by experts.

(ISONO Yayoi)

Essay 2   Industrial Parks and Worker Health Damage

In Thailand pollution by agro-industry and by small and medium-sized enterprises is especially serious. Meanwhile, as industrial parks are built throughout the country to help develop industry and farming areas, factory workers are suffering from grave health damage caused by the large factories of joint enterprises. A case in point is a factory in a Lampang Province industrial park.

Built in 1972, this industrial park has about 20 electronics plants. After the park's construction 14 workers died, 12 of whom worked in electronics factories. The other two were the two-year-old child and the three-month-old infant of women workers who had died. Workers began suspecting that hazardous substances from the factories were the cause. Although there was no unmistakable record of hazardous substance accumulation, postmortem examinations suggest their deaths were caused by heavy metal contamination. Nevertheless, the deaths were officially explained away as AIDS-related. Workers are suspicious because they are directly exposed to hazardous substances and forced to work with no consideration whatsoever for their working environment, and because many people working there still become ill and have no choice but to resign. In one case, a woman worker employed by Electro Ceramics, a Japanese-Thai joint venture, suffered grave health damage and filed a lawsuit against her factory demanding 6 million baht in compensation. This woman, whose job involved wiping away alumina powder with her fingers, had symptoms including headaches, nausea, numbness, and even loss of balance. Her affliction was determined to be alumina poisoning, and she underwent medical treatment but recovered only partially. Cutting-edge industries gathered in an industrial park in Thailand's interior seeking cheap labor from the surrounding rural area, and as with this woman, many of the workers are in their 20s and 30s. In Japan female workers handling hazardous substances would be subject to strict controls, but at this industrial park women work long hours under poor conditions for low pay. These companies do not recognize the causal relationship between health damage and poor working conditions. Health studies should be conducted at existing industrial parks, and, when siting a new park, nearby citizens should be provided with information and sufficient opportunity to participate.

(Uthaiwan KANCHANAKAMOL)

Chapter 4
Malaysia

1. Introduction

Formerly Malaysia was characterized by rubber plantations and tin mines, and by lush tropical jungles inhabited by orangutans. In the production of rubber and tin, Malaysia is still in the running for first or second place in the world, but those lush tropical forests have largely disappeared. Now the country is one of the world's major manufacturers of consumer electronics, and its economy has changed to the point where industrial goods account for most exports.

People traveling certain roads from Penang International Airport toward downtown Penang will see long lines of foreign-capital factories on both sides. This is the Bayan Lepas Industrial Park. Many are Japanese companies, but there are also many signs with German and U.S. company names. Crossing Penang Bridge to the Malay Peninsula, ocean pollution becomes more evident as one approaches the peninsula. Nearby is Prai Industrial Park, one of Malaysia's largest, where the factories of Japanese corporations have a conspicuous presence.

Since the latter half of the 1960s, Malaysia has brought in much foreign capital and established many export processing zones, thereby achieving fast industrialization, which in turn induced fast urbanization that has resulted in heavy traffic jams clogging the streets of Kuala Lumpur, one of Southeast Asia's most beautiful cities. And since about 1980 the excessive logging of tropical timber in eastern Malaysia has caused the loss of forests.

This chapter will focus on a number of points in briefly examining how environmental damage in Malaysia began, how it changed in the subsequent process of development, how people began calling for environmental protection, how their demands were reflected in actual policy, and how government administration and legislation responded.

2. Environmental Damage from Colonial Times

2-1 Tin Mining and Water Pollution

Although Portugal had first colonized part of Malaysia, it was in the latter half of the 19th century that colonization geared up when England made Penang Island into a free trade zone. England had noted the value of the high-quality tin mines throughout Malaysia, and at the beginning of the 20th century put major effort into tin mine development. Much tin mining was pit mining that resulted in many open pit mines in tin-producing areas such as Kuala Lumpur and Ipoh. Mining waste turned to mud and polluted many rivers, and that was the outset of environmental damage in Malaysia.

2-2 Plantations and Deforestation

Worse than tin mining in terms of diminishing species diversity and forest cover was the creation of large-scale plantations for rubber and other products. In response to rubber demand, which grew quickly with the advent of the 20th century, the English planted rubber tree plantations over much of the Malay Peninsula, increasing the mere 4,500 ha existing in 1903 to 810,000 ha in 1921. The ecosystem was devastated and natural forests disappeared. Rubber profited the English handsomely, but the monoculture plantations deprived Malaysia of biodiversity and have deeply scarred the land in many places.

2-3 Land Ownership System and Modern Values

Huge rubber plantations were made possible by the highly simplified "modern" land ownership system imported by the English. Formerly in Malaysia the concept of land ownership had not been necessarily well-defined, and there had been a variety of customary land-use relationships, but Western influence changed the system to make capitalistic land use possible, which deprived the Malaysians of their land. The English simplified relationships of land use and ownership to "modernize" the system by making land the possession of the sultan, who was Malaysia's nominal ruler, and then had the sultan award exclusive land use rights to European entrepreneurs. By changing the land use system and creating a rubber production base with plantations, Western countries integrated Malaysia into the world capitalist economy and made it into the world's largest rubber producer. Integration also brought "modern" Western values and knowledge, which heavily influenced local elites. This also did away with the indigenous traditional land use system, techniques, and socioeconomic system. Some have observed that this is one cause of present environmental damage in Malaysia.

3. Modern Industrialization and Environmental Damage

3-1 Foreign Capital Brings Industrialization and Pollution

• Rapid Industrialization

Rapid industrialization began in the latter half of the 1960s. While industrial production accounted for only 8.7% of GNP in 1960, by 1993 that expanded 70-fold, and accounted for 30.2% of GNP. One national policy was industrialization using foreign capital, and Japanese capital too flowed into Malaysia primarily in two waves, one in the mid-1970s and the other in the latter half of the 1980s. Malaysia is now a major production base for Japan's electrical appliance industry, whose televisions, air conditioners, and other products are exported to Japan and other industrialized countries. This rapid industrialization, however, has caused problems including heavy industrial pollution, concentration of the population in cities, and urban pollution.

• Industrial Pollution in Penang State

An early pollution problem was the effluent, smoke, and other pollutants from an industrial park near Penang Island (Penang State), which is both one of Malaysia's finest tourism areas and its oldest commercial district. In 1970 Penang State had industry consisting of only 210 ha and 31 factories, which grew swiftly to 948 ha and 706 factories in 1995.

Prai Industrial Park, across the water from Penang Island, is one of the oldest parks and has many Japanese companies, some of which have been operating since the mid-1970s. Sewage from this industrial park ruined the fishing for a fishing village south of the park, whose inhabitants thereby lost their livelihood. Villagers say that in 1975, several years after the factories had started operating, the fish catch dropped to one-fifth that of before, and the disagreeable smell of the fish made them unfit not only for market but even for domestic consumption.

In 1977 a Japanese newspaper correspondent went to see the effluent gate behind the industrial park, where the emerging sewage produced white bubbles and gave off a stench. An investigation by the Consumers Association of Penang revealed the presence of heavy metals such as cadmium and mercury. Although it was hard to say which companies were causing the pollution, the effluent from Japanese dyeing companies was suspect, and this became an issue of "pollution export" from Japan. Effluent from Prai continues to contaminate the surrounding sea. Several surveys performed from the latter half of the 1980s through the 1990s showed that contamination by lead and zinc is spreading in the nearby sea, and another survey revealed that oyster growth in the polluted area is very poor.

3-2 The ARE Incident

• Pollution Export from Japan

About 100 km south of Penang is Ipoh, Malaysia's third-largest city. Situated in the middle of Kinta Valley, Malaysia's largest tin-producing district, it is an area with many ethnic Chinese. Here, about 8 km south of Ipoh, the small village of Bukit Merah on the Lahat Highway was founded in the early 1950s by mainly tin miners. A small industrial park was established by the village, and around 1980 a company called Asian Rare Earth Sdn. Bhd. (ARE) built a factory here, which began operating in 1982. A major issue arose when people claimed that the factory was sloppy in its use of radioactive materials. ARE was a joint venture 35% owned by a Japanese Company, the then Mitsubishi Chemical Industries.

The plant discarded its waste in a pond behind the factory, and villagers were not informed about the materials processed there. They passed near the dump, their children played there, and some people even used the waste as "fertilizer" in accordance with advice from factory people. That waste, however, contained radioactive thorium. ARE's work was extracting and refining rare earths from monazite, a substance from tin tailings, which were used to produce the red color in TV picture tubes, small magnets for cassette recorders, petroleum catalysts, and other products that were almost all exported to Japan. Those processes yielded the radioactive substance thorium hydroxide. ARE should have strictly controlled it, but at first discarded it behind the factory with hardly any controls, something that would have been unthinkable in Japan. Factories in Japan too once extracted rare earths from monazite, but restrictions were strengthened during the groundswell in the anti-pollution movement of the 1960s, and strict controls were instituted under the 1968 Nuclear Reactor Regulation Law. So it was that after 1971 the process of extracting rare earths from monazite disappeared from Japan. Mitsubishi Chemical Industries was well aware of this sequence of events, and also no doubt knew the seriousness of its liability as a company using hazardous substances because its liability had been questioned, along with siting negligence, by the 1972 decision in the Yokkaichi Pollution Lawsuit. In 1973 it planned the Malaysian venture and in 1979 founded ARE. Given the number of years that had elapsed, Mitsubishi should naturally have enacted strict measures and conducted an EIA in advance, especially because within a 1 km radius of the ARE plant were the 10,000 people of Bukit Merah, and a new housing development of about 3,000 people, as well as a stream flowing behind the factory. The situation called for an EIA especially in view of Mitsubishi's bitter experience at Yokkaichi, but it did not perform one. Instead, it began operations without building a storage facility even while aware that radioactive wastes were produced.

• Health Damage and Local Anger

This slipshod waste management allowed the radioactive waste to be spread around the area. In 1984 Professor Ichikawa Sadao of Saitama University in Japan performed a survey around the ARE factory at the request of the local citizens. Around the pond behind the factory he discovered high readings of 7-48 times natural radiation, with the highest reading detected being over 700 times. A subsequent survey discovered high radiation in other places. Radioactive substances made their way into local citizens' bodies via air and food. Around 1987 decreased leukocyte counts were found in village children, there was a very high abnormal birth rate, and lead concentration in blood was increasing. Then in 1988 and 1989 health problems started afflicting many children. A number of children contracted leukemia, and two of them are already dead. There were also cases of child cancer and congenital disorders, a high incidence rate for a village of 10,000 people.

Locals were greatly angered by the factory for causing this state of affairs, and in 1987 they made frequent demonstration marches on the factory, but the factory closed its gates and would not listen to them.

• Pollution Lawsuit and Factory Closure

Locals filed a lawsuit to shut down the factory, and while the case was in court there was an attempt at repression in which leaders of the citizens' movement, activists of supporting environmental groups, and lawyers were suddenly arrested, but citizens carried on their campaign in which over 2,000 people gathered at the courthouse for each hearing. This dedication to the cause evoked international interest so that experts from Japan, Canada, the U.S., and other countries came to Malaysia each time the case was heard. Hearings concluded in the spring of 1990, and when the decision finally came two years later it awarded victory to the plaintiffs, recognizing the pollution and the plaintiffs' demand for factory closure. The decision was big news not only in Malaysia, but also in Japan where it was called the first case that passed judgment on pollution export. But then about 18 months later, in an appeal to Malaysia's Supreme Court, a reversal on the decision handed defeat to the plaintiffs with hardly any convincing reason. Nevertheless, under pressure of world public opinion, Mitsubishi Chemical Industries and ARE announced in January 1994 that they would close the plant.

This case is considered a typical example of pollution export especially by a partly Japanese-owned company, and the lesson is that even though companies may have pollution abatement technologies, they do not necessarily make use of them. The guidelines released in April 1990 by Japan's Federation of Economic Organizations, which were occasioned in part by this case, ask Japanese companies to apply the most stringent controls on hazardous substances everywhere in the world. While this is only natural, the unfortunate fact is that currently it can only happen if there is some sort of monitoring.

3-3 Spotlight on Industrial Wastes

While the ARE case is an extreme one, Malaysia has long lacked fundamental measures on how to manage industrial wastes. Though industrialization has proceeded since the latter half of the 1960s, there were hardly any measures to deal with the resulting industrial wastes, and no disposal sites for them, so that basically each factory would keep the wastes on its own site. But this situation became a problem in the early 1980s, and some studies were performed with the help of Australian consultants and U.S. aid. Recent data show that as of 1994 Malaysian industry generated about 420,000 cubic meters of industrial wastes (i.e., designated hazardous wastes) annually, and it was evident that factories could no longer keep this much on site. Thus in 1985 the Malaysian government's Department of Environment (DOE) surveyed candidates for industrial waste disposal sites, and chose the Bukit Nenas site in Negiri Sembilan State. As this is a headwaters area, local citizens and environmental groups mounted an opposition movement that was pushed aside partly because this was a national project under the DOE. The EIA qualified and construction finished in 1996, with operation beginning in January 1997. This site has facilities for both landfilling and incineration, and five years after beginning operation it will be able to handle 300,000 tons of wastes annually.

However, environmental groups are criticizing the disposal of hazardous wastes by landfilling and incineration. Negative environmental impacts will always arise whether wastes are burned or buried, and there are concerns of such impacts at Bukit Nenas as well. Even some people in the DOE say that waste generators should seriously consider plans to minimize waste, and alternatives such as clean technologies. There are those in Japanese industry, which has invested much in Malaysia, who see Malaysia's industrial waste problem as a confining factor, but as businesses now operate globally, they should welcome demands for tight waste management as something that will occasion the development of new technologies.

4. Resort and Golf Course Development

4-1 Penang Hill Development Issue

One more characteristic of current development in Malaysia is tourism development of resorts, golf courses, and the like, and the construction of large-scale housing projects for the elderly citizens of developed countries. Penang Hill is a representative example.

Penang Hill is a prominence in central Penang Island that peaks at 830 meters. It is valuable as the island's water source and the variety of ecosystems remaining there, and as a valuable historical and cultural area for its old buildings dating from the colonial era. Currently there is no road up the hill, so the only access is by a cable car that is both transportation for area residents and the only way for tourists to ascend.

A plan to make Penang Hill into a giant resort surfaced suddenly in September 1990 in the form of a memorandum between the Penang State government and a subsidiary of the Berjaya financial combine. It called for the development of 900 acres obtained by leveling the top of Penang Hill, and included a botanical garden, two giant hotels, a shopping center and sports center, a huge mall with a theater and other facilities, a golf course, a mammoth amusement park including a spook house and roller coaster, a residential area for the elderly, and more. Many Penang residents and other Malaysians were astonished at this plan and formed an organization called "Friends of Penang Hill," whose members included a large variety of organizations. The opposition movement, whose core was "Friends of Penang Hill," grew to such proportions that on January 24, 1992 it forced the Malaysian DOE to refuse the environmental impact assessment of the development plan, which also forced the Penang State government to issue a statement saying the plan had been abandoned.

This case is the first instance of an EIA with actual citizen participation under the Environmental Quality Act, and it also stimulated the concern of many people toward environmental protection. Another reason for the close interest in this case is that the citizens had a say in local government, which resulted in replacement of the Chief Minister with a more cautious person.

4-2 The Tragedy of Langkawi

But it is not always possible to stop resort development projects such as Penang Hill. A considerable portion of Langkawi Island, about 100 km north of Penang, is devastated because of tourism development. In the latter half of the 1980s Langkawi came into the spotlight as a tourism location to replace Penang Island, which had been industrialized and polluted. Langkawi's northern area had been a place of picturesque beaches, but owing to the withdrawal of resort developers in the middle of the project, the area turned into a pathetic scene resembling a trash dump with its land stripped of vegetation and punctuated here and there by pools of water. There was also construction of hotels and golf courses financed by Japanese capital, but there was too little consideration for conserving the island's ecosystem. In particular the coastline was affected by hotel construction, but no one had seen fit to consult with the fisherfolk, who had been living there a long time. In addition, Langkawi was designated a free port, bringing a huge influx of foreign capital. The upshot was, ironically, that the island's natural beauty brought about its own demise, which undermined the island residents' means of subsistence.

4-3 Golf Courses

In June 1991 the official publication of the Consumers Association of Penang carried a report on golf course development in Malaysia. In 1974 the country had had only 45 golf courses, increasing to 72 in 1990, with 80 more plans either in progress or about to begin. Since not a few golf courses have high proportions of Japanese golfers, there is criticism that such rapid golf course development is meant for Japanese people, and there is even criticism that, owing to the cessation of golf course development in Japan after the economic bubble burst, there is an attempt to "export" golf course development in the manner of "exporting" pollution.

4-4 Highland Highway Plan

Intent on tourism development, Malaysia named both 1990 and 1994 "Malaysia Tourism Year." There is talk that even the abandoned Penang Hill development scheme will reappear in a different guise. One astonishing plan is to build a "Highland Highway" across the peaks of the mountain chain running through the middle of the Malay Peninsula, which would consist in an expressway linking the highland resorts in the mountains. Malaysia has experienced frequent landslides in recent years, and reckless development is cited as a cause, so grave environmental damage is foreseen if this road is built through the mountains. Malaysian environmental groups therefore strongly oppose this plan.

5. Responses to Environmental Problems

As the foregoing discussion has shown, the environment in Malaysia has been confronted with tin mining and rubber plantations from the colonial period, and now with industrialization and tourism development. Below we shall examine how Malaysia has dealt with these problems.

5-1 Environmental Law and Administration

• From Individual Laws to the Environmental Quality Act

For a long time Malaysia had no basic environmental law, instead coping with environmental problems using individual laws that were originally conceived for purposes other than protecting the environment, such as laws covering bodies of water, mining, forests, and the like. But this was poorly suited to comprehensive government administration of the environment because the laws were meant for special purposes, or authority was divided among several government agencies. So to integrate environmental law Malaysia in 1974 passed the Environmental Quality Act, a sweeping act that establishes various powers to regulate pollution, calls for creating an EIA system, and attempts to integrate measures for environmental consideration when implementing development plans. Because it takes the form of a basic law, the regulations, decrees, and other injunctions based upon it are virtual law. It also establishes the office of the Director General of Environmental Quality, who is given powerful authority in issuing regulations. In 1975 the DOE was created as the first administrative organization for the environment. Although it is formally under the Ministry of Science, Technology and Environment, it has a strongly independent character and has played its own unique part in environmental protection.

• Subsequent Developments

Various regulations were based on the Environmental Quality Act, but all are individual pollution control regulations except for the 1987 EIA regulation, and despite the initial intent, the law has actually not managed to prevent pollution or to integrate environmental measures in plan implementation. Even individual regulations for pollution prevention have yet to show concrete results, as with the worsening marine pollution in the Penang area. In 1992 the law was thoroughly reassessed, but this has led to no legislative action, and the law's problems remain unsolved.

• Authority Conflicts with States

Environment-related administration in Malaysia also involves the problem of federal and state authority because the Constitution gives states authority over land, forests, water sources, fisheries, and agriculture. Thus integrated government action on the environment absolutely requires coordination of federal and state authority.

5-2 EIA System

• EIA Order and Handbook

Article 34 of the Environmental Quality Act provided for Malaysia's EIA system, but it actually assumed concrete form under the EIA order of 1987. Implemented in April 1988, the order merely listed the affected project types, while specific EIA procedures are spelled out in guidelines called a "handbook." First issued in June 1987 and revised in August 1995, the handbook forms an integral whole with the Environmental Quality Act and the EIA order, and is legally binding. It sets forth the following points as the seven important items for EIA procedures: (1) Clearly state the project's necessity, (2) consider alternatives (alternatives in siting and design, as well as the "no project" alternative), (3) consider mitigation, (4) gather environmental information, (5) provide for local citizen participation, (6) perform a cost/benefit analysis, and (7) carry out monitoring. It also calls for EIA performance at the earliest possible date.

• Pioneering Role at Penang Hill

It was at Penang Hill where Malaysia's EIA system actually played a major role. As this scheme involved large-scale development and attracted the concern of many people early on, the Penang state government initially announced that it would carry out an assessment on the development plan as a whole. Here the procedures' flexibility proved effective because the plan's inadequacies were pointed out at the draft assessment stage, so that many people who had read the draft expressed their opinions at the final assessment stage. With as many as 1,400 opinions submitted, this was the first instance of true citizen participation in EIA procedures. In response to these opinions the federal government established a screening committee to examine the final assessment. Such committees are legal entities defined by the law and handbook, and a different one is organized for each case. The committee for Penang Hill included not only members from various government departments, but also members of "Friends of Penang Hill," making for unprecedented citizen participation at the screening committee level. Committee rejection of the inadequate assessment document led to abandoning of the plan, resulting in a success story about a comprehensive assessment performed early and with real citizen participation. Although carrying out EIA procedures is nearly optional in Malaysia, the handbook is advanced because it calls for early implementation, submission of alternatives, establishing a screening committee, and the like. With the legislation of related laws, this could very well become an effective means of integrating environmental measures into development plans.

• Backsliding at Bakun

Later, however, the Bakun Dam planned in the eastern Malaysian state of Sarawak, served to show how EIAs were losing their effectiveness as a means of environmental protection. This dam had been planned in the 1980s, but was canceled in 1986 because its environmental impact was so large, then in 1993 it suddenly reappeared. Its main purpose was to produce hydropower that would be transmitted by means of an undersea cable to the Malay Peninsula, a huge project that would inundate 70,000 ha and affect about 10,000 indigenous inhabitants. Problems included the way in which the project was decided, such as placing orders with companies that have no dam construction experience. Here too EIA procedures were initiated in accordance with the Environmental Quality Act, the 1987 decree, and the handbook, but suddenly in 1994 Sarawak established EIA procedures that excluded local citizen participation, screening committees, and consideration of social impacts on native peoples. Then in 1995 the federal government revised the EIA decree to exempt hydropower projects in Sarawak, and there were even attempts to make this retroactive. This was none other than an attempt to pursue the Bakun dam project by taking advantage of the constitutional mixing of federal and state powers. Indigenous inhabitants claimed this was illegal and filed a lawsuit seeking: confirmation that the 1995 decree revision was illegal, EIA implementation in accordance with federal procedures, and an injunction on the project to date. As this lawsuit called the legality of federal law into question, the first hearing was in the Kuala Lumpur High Court, which on June 19, 1996 almost totally recognized the citizens' claims and issued an injunction stopping the project. However, the government and dam contractors immediately filed an appeal, and ultimately the decision was reversed. This case shows that as Malaysia's EIA procedures are gradually established, there are attempts to weaken the system by using state authority in a way contrary to its original purpose, which now seriously challenges conservation efforts.

5-3 Citizen Awareness and the Role of NGOs

In Malaysia there has not always been such great concern for environmental protection among the citizens, but they have gradually become more concerned as their own livelihoods are threatened by grave industrial pollution and damage to their immediate environments. A feature of Malaysia's environmental movement is its very active NGOs. Since the days of the pollution at Prai Industrial Park to the ARE incident and the Penang Hill development issue, the Consumers Association of Penang, Friends of the Earth Malaysia, and other NGOs have always worked to educate the people about the environment, conducted scientific studies, supplied legal support, and even provided expert technology. This chapter has dealt chiefly with the situation on the peninsula, but NGOs have played a major role also with regard to logging issues in eastern Malaysia. Further, in organizing broad citizens' movements, NGOs do much to raise the environmental consciousness of local citizens. The Penang Hill experience is a good example because the coalition of varied groups and people engendered political reform at the Penang State government level, which served to elicit the interest of all Malaysian citizens in protecting the environment. What is more, the world focused on the international role that Malaysian NGOs played at the Earth Summit, held about a half year after the Penang Hill problem was more or less resolved in early 1992. At the Earth Summit, Martin Khor and Chee Yoke Ling of the Third World Network, Gurmit Singh of the Environmental Protection Society of Malaysia, and others had considerable influence on not only NGO efforts, but also government-level efforts. The Ipoh High Court decision on the ARE case that came a short time later demonstrated the importance of solidarity between Malaysian and international NGOs. Heretofore transnational corporations have caused a variety of problems because their actions are not watched as closely in other countries as in their own, but here too the international NGO network used the tools of the information age to its advantage, resulting in Japanese media coverage of ARE's actions in Malaysia, which led to NGO initiatives and Dietmember efforts in Japan, ultimately affecting even the actions of industry and the government. Here is an event illustrating specifically how campaigns by NGOs and citizens across national borders can stop environmental damage occurring across national borders.

6. Summation

As the Highland Highway example shows, Malaysia with its continuing rapid economic growth still faces a crisis of grave environmental damage, and there is ongoing repression of those who oppose such environmentally destructive development. Yet, the Penang Hill campaign demonstrated that citizens and NGOs are beginning to develop self-confidence in their own capacity to conserve the environment. As people increasingly join hands across national borders, even Malaysia's government has begun to note the importance of the environment. Further, owing to problems associated with the resource development and plantations for export crops existing since premodern times, present environmental damage in Malaysia and other developing countries — from modern industrial and tourism development to the logging that this chapter could not cover in detail — is often closely tied to the wasteful economies of the developed countries.

(KOJIMA Nobuo, with the cooperation of MEENAKSHI Raman and the Consumers' Association of Penang)

Chapter 5
Republic of Indonesia

1. Introduction: Indonesia's Natural Environment and Economy

1-1 Two Ecosystems

The Republic of Indonesia is a nation of islands lying over the equator and extending 5,200 km east to west and 1,900 km north to south. Java especially has many volcanoes whose ash has contributed to its fertile soil. Wet rice agriculture prevailed on this heavily populated island and its surrounding region, called "Inner Indonesia." The region consisting of Kalimantan, Sumatra, Sulawesi, the Moluccas, Irian Jaya, and other islands, called "Outer Indonesia," is characterized by poor soil and therefore mostly swidden agriculture.

The American anthropologist Clifford Geertz had a major impact on Indonesian studies when he observed that Indonesia has two contrasting ecosystems: The rice paddy ecosystem of Inner Indonesia and the swidden ecosystem of Outer Indonesia.[1]

Some characteristics of the rice paddy ecosystem are monoculture cultivation of rice, rice paddy scenery differing totally from that of tropical forests, stable sustainability due to nutrient-carrying water, dependence on irrigation channels, and high population carrying capacity. By contrast, the swidden ecosystem is characterized by cultivation of multiple crops (dry rice, tubers, grains other than rice, vegetables), swiddens structured like tropical forests, a delicate balance resulting from the cycling of nutrients between plants (i.e., trees and crops) and poor soil, and low carrying capacity.

1-2 Uneven Population Distribution

Indonesia's overall population density in 1995 was 101 persons per square km, which is not especially dense, but far more of the population is on Java, with its fertile soil and rice paddy ecosystem. While Java accounts for only 7% of Indonesia's land area, it hosts over 115 million people, or 60% of the population, making for the high population density of 900 people per square km. Bali's density is 514, but the other areas and islands have low densities. At the same time, some provinces have uneven internal distributions.

1-3 Forests by Region

The 1994 "Agreement on Forest Utilization Plans" (TGHK) says that Indonesia has 140.4 million ha of forested land, of which 113.8 million ha are to be maintained as forest throughout the future, and of which 92.4 million ha are now forested. Using either figure, Indonesia has the world's second-largest tropical forest area after Brazil. But the area being deforested is also large. An FAO report[2] estimates that from 1981 to 1990 Indonesia lost 1.2 million ha of forest annually, a figure that is 8% of the total 15.4 million ha of world forest loss. Indonesia's statistical yearbook *Statistik Indonesia* says that about 60% of the country is forested. But while Kalimantan, Irian Jaya, and the Moluccas are highly forested at 68% (Table 1), Java's proportion is only 23%. Indonesia's per capita forest area is 0.6 ha, but here too there are large regional differences, ranging from the 17.5 ha on Irian Jaya to the 0.3 ha in the Nusa Tenggara. Per capita forest area on Java is near zero.

1-4 Economic Turning Point

Statistics show that Indonesians' dependence on the primary industries has declined as those industries' proportion of the GDP has decreased from 50% in 1965 to 40% in 1973.

Since Indonesia's independence its government has pursued economic development mainly through industrialization to substitute for imports, and to that end it has held to a "full set" industrialization policy that attempts to build production systems for all products from consumer goods to their parts, intermediate inputs, machinery, and other production equip-

---

[1] Geertz, C. *Agricultural Involution: The Processes of Ecological Change in Indonesia*, Univ. of California Press, 1963.

[2] "Forest Resources Assessment 1990: Tropical Countries," FAO Forestry Paper No. 112, 1993.

Table 1 Indonesia's Regional Characteristics

Region	Soil	Ecosystem	Main food	Population	Migration	Forested proportion	Per capita forested area	Per capita non-oil GDP	Growth rate of non-oil GDP
Java	Fertile	Rice paddies	Wet rice	Dense	Outflux	Low	Nearly zero		Low
Nusa Tenggara	Poor	Swidden	Other grains					Low	
Sumatra	Poor	Swidden	Dry rice						
Kalimantan	Poor	Swidden	Dry rice	Sparse	Influx	High	Large	High	High
Sulawesi	Poor	Swidden	Tubers, other grains, sago						
Maluku	Poor	Swidden	Tubers, other grains, sago			High			
Irian Jaya	Poor	Swidden	Tubers	Sparse		High	Large		High

Note: Though one of the Nusa Tenggara, Bali is actually similar to Java in many indicators.

ment. But in the early 1980s there were decreased prospects for income from oil and gas exports, which had supported this policy of dependence on internal demand, and the country directed greater expectations toward its industrial sector. In 1983 Indonesia began working on a structural adjustment and came up with a series of deregulation measures; the government instituted a policy of exporting more highly processed goods, weakened restrictions on foreign currency, and lowered the exchange rate. The manufacturing industries' share of GDP rose from 8% in 1965 to 10% in 1973, 11% in 1983, then to 21% in 1992, finally surpassing agriculture's 19%, then moving on to 24% in 1994. This brought Indonesia's domestic industrial structure to a turning point. Advancing industrialization caused pollution problems mainly in the urbanized areas of Java, while deforestation became serious in rural areas like Kalimantan.

This chapter will focus on pollution and deforestation, two major elements of Indonesia's environmental problems, and discuss the roles of citizens' movements and NGOs in tackling them.

2. Pollution and Remedial Measures

Indonesia experienced an influx of direct foreign investment owing to the high valuation of the yen and NIEs currencies in the latter half of the 1980s, and then around 1990 Indonesia's economy started enjoying prosperity, mainly in the manufacturing industries, which, except for the oil- and gas-related industries, have maintained growth of at least 10% a year. In paper/pulp, textiles, chemicals, electronics, and many other areas the manufacturing sector enjoys growing production (Table 2). And even though the extractive sector shows no rise in crude oil production, there are increases in the production of coal, copper, and other items. This very rapid growth by the manufacturing industries is causing environmental problems, and this is especially the case with paper and pulp, as we shall see in the following section.

After discussing the current state of pollution problems, we shall touch upon the kinds of pollution disputes that arose in the first half of this decade, then cover government policy

Table 2 Production Volume and Annual Growth of Main Products in Indonesia

		Production			Annual growth rate (%)	
	Units	1983	1988	1993	83-88	88-93
Manufacturing Industries						
Textile thread	mil m	2347.2	3503.0	7878.5	8.3	17.6
Cloth	1,000 bales	1370.0	2712.3	4933.7	14.6	12.7
Plywood	1,000 m³	2.6	6.9	9.5	21.6	6.6
Paper	1,000 t	369.2	948.2	2489.3	20.8	21.3
Pulp	1,000 t		103.7	1304.6	-	65.9
Caustic soda	1,000 t	14.4	38.0	362.1	21.4	57.0
Acetylene	1,000 m³	1504.0	1985.0	6344.0	5.7	26.2
Sulfuric acid	1,000 t	224.0	875.7	1000.1	31.3	2.7
Monosodium glutamate	1,000 t	33.4	64.6	148.6	14.1	18.1
Plate glass	1,000 t	146.8	312.6	429.0	16.3	6.5
Sponge iron	1,000 t	541.0	984.8	1428.6	12.7	7.7
Aluminum ingots	1,000 t	115.0	199.0	192.6	11.6	0.7
Televisions	1,000 units	622.8	521.9	1476.0	-3.5	23.1
Storage batteries	millions	4.1	6.1	11.6	8.3	13.7
Dry cells	millions	633.6	1016.0	1463.4	9.9	7.6
Urea fertilizer	1,000 t	2255.0	4245.9	5132.7	13.5	3.9
Ammonium sulfate fertilizer	1,000 t	208.0	586.1	529.6	23.0	-2.0
Portland cement	1,000 t	8102.1	13218.0	18990.0	10.3	7.5
Rubber/canvas shoes	millions	28.4	44.6	350.1	9.4	51.0
Vehicle assembly	1,000 units	155.7	166.7	209.2	1.4	4.6
Extractive Industries						
Crude oil	millions of bbl	477.9	435.2	559.9	-1.9	5.2
Coal	1,000 t	410.5	5175.7	28559.5	66.0	40.7
Copper	1,000 t	199.7	302.7	960.0	8.7	26.0
Other						
Primary energy consumption	1,000 bbl oil equiv.	223613.0	300915.0	426250.0	6.1	7.2
1983-base real GDP	%	100.0	128.7	179.9	5.2	6.9

Sources: Prepared from various years' editions of the *Lamprian: Pidato Kenegaraan Presiden Republik Indonesia* and from national income statistics provided by the Central Bureau of Statistics.

on pollution, especially that on cleaning up rivers, which is the most highly developed.

2-1 Worsening Pollution

• Water

A problem that received early attention in Indonesia is marine pollution by oil spills from tankers passing through the Strait of Malacca. Since an accident by the *Showa-maru* in 1975, there have been a number of major accidents. In a September 1992 collision the *Nagasaki Spirit*, a tanker of Libyan registry, damaged local fishing when it spilled between several thousand and 10,000 tons of crude oil, and in a January 1993 collision a Danish-owned tanker spilled over 27,000 tons of crude.

Heavy metal contamination is serious in Jakarta and Surabaya bays. In Jakarta this was revealed when the National Atomic Energy Agency and the Indonesia Institute of Sciences conducted a survey in 1977-78, which showed high concentrations of mercury and other heavy metals in sea water and organisms such as shrimp. The City Urban and Environment Research Center reports that in recent years the concentrations have risen further.[3] Graywater flowing into Jakarta Bay turns the water brown, and trash floats on the water. About 70% of the Seribu Islands fisherfolk who used to make a living from Jakarta Bay have reportedly given up fishing and moved to the cities.[4]

With the urban population increasing despite the lack of sewerage and other infrastructure, graywater inflow is worsening river water quality. For example, in 1993 at least 14% of Jakarta households were using toilets without septic tanks, or were simply using ponds or rivers for the purpose. Even toilets with septic tanks are not properly maintained, and often the sludge is not removed. And even if sludge is pumped out, many cities have no sludge treatment facilities, so untreated sludge is dumped into rivers and the like. A pilot project for Jakarta's sewage has just begun and covers only about 3% of the city's area.

Expansion of the manufacturing industries since the latter half of the 1980s has brought about pollution disputes around the country (Table 3). Many of these problems caused by the manufacturing and mining sectors concern water, which is perhaps because, owing to its use in daily life and in certain applications such as in rice paddies and aquaculture ponds, problems quickly become apparent. Especially the pulp and paper industry has rapidly increased production since the latter half of the 1980s, triggering disputes around the country.

• Air

Disputes arise between citizens and factories over air pollution as well, and include problems such as an ammonia leak from fertilizer factories and smoke from cement plants. There are also reports of widespread, high-concentration airborne particulates and lead in Jakarta and other cities. The main causes of airborne particulates are motor vehicles, the use of firewood in factories and homes, and open-air trash burning, while lead comes from leaded gasoline. Health damage is no doubt quite severe. One estimate,[5] for example, says that in 1990 air pollution by airborne particulates and lead caused about 1 trillion rupiah (about US$500 million) in health damage.

• Wastes

In many Indonesian cities such as Jakarta, inadequate trash collection is a cause of water and air pollution. Uncollected trash is discarded in rivers or burned on riverbanks. *Environmental Statistics of Indonesia*, issued by Indonesia's Central Bureau of Statistics, estimates that in Jakarta 25,404 cubic meters of trash were discarded daily in 1993. Over 21,384 cubic meters were hauled away for proper disposal, while the rest is thought to have been recycled, burned on riverbanks, or otherwise disposed. To deal with industrial wastes, Indonesia in 1994 built Southeast Asia's first treatment and disposal center in West Java, and it is already operating.

2-2 Coming to Terms with Pollution

It was in the 1970s when Indonesia's government first passed legislation to deal with pollution. In the latter half of the 1970s laws were passed on the provincial level, such as in Jakarta and East Java, where river pollution was serious, and in 1982 Indonesia passed the Environmental Management Act. But these were toothless laws. Catalyzed by more environmentally related aid and the increasing gravity of environmental problems in conjunction with economic development beginning in the latter half of the 1980s, government measures to control pollution gradually became more effective beginning about 1990. Nevertheless, this does not mean that effective pollution measures have been implemented throughout Indonesia and in all sectors. Here we shall examine advances in government action on pollution and the associated problems through the lens of measures to clean up rivers, which are most advanced.[6]

PROKASIH is a river clean-up program started in 1989. Coordinated by the then Ministry of State for Population and Environment and with the participation of eight of Indonesia's 27 provinces, it is an attempt to decrease the amounts of pollutants from companies in the watersheds of the 20 rivers covered. Currently the Environmental Impact Management Agency, which was established to implement environmental controls at the national level,

---

3   *Jakarta Post*, September 2, 1991.
4   *Jakarta Post*, January 6, 1996.

5   In 1992 prices. Haryoto Kusnoputranto, "Urban Protection and Pollution Control in Jabotabek: Impacts on Health, Economy, and the Environment," paper presented at an August 1995 seminar, Issues of Using Leaded and Unleaded Gasoline: Impacts on Health and the Environment.
6   Jay Nagendran, *PROKASIH: A River Cleanup Program in Indonesia, Environmental Management Development in Indonesia Project*, Jakarta and Halifax, 1991. Environmental Impact Management Agency, *PROKASIH: Evaluating the Last Four Years and Looking to the Future*, Jakarta, 1994.

Table 3 Major Pollution Disputes and Incidents, 1990-1995

Companies	Industry/products	Area	Description
Eight companies including PT. SDC, PT. KIIS, PT. PD, PT. APTI, and PT. SM	Chemical, paper, soap, textiles, foods, etc.	Semarang, Central Java, Tapak River	Damage to paddies, aquaculture ponds, etc. from 1978; 15 NGOs start boycott in 1991, settlement reached that year.
PPD	Buses	Jakarta, Mookevart River	River polluted by discarding lubricating oil, parts, etc. Jakarta Province governor warns PPD.
PT. TI	Candy	Jakarta	Environment minister points out that PT. IT is the most polluting company of those subject to the Clean River Program. In 1992 the company shut down for six months and invested in pollution control.
PT. Indah Kiat	Pulp	Riau	Water pollution. Citizens complain to Environment Ministry in June 1990.
PT. SME, PT. MK, PT. IF, PT. SPM, PT. PA	Textiles, chemicals, pharmaceuticals, rubber	Bukasi, West Java, Sadang River	Companies start operating about 1989. Two children die after falling into polluted river.
PT. Indra Era	Lead recovery from old batteries	Pasuruan, East Java	Began operating in 1989. Lead contamination.
About 200 tapioca factories	Tapioca processing	Central Java	Contaminated about 1,000 ha of rice paddies and shrimp aquaculture ponds.
PTP XVIII	Rubber	South Kalimantan	Rubber effluent contaminated about 2,000 ha of fields and rice paddies.
PT. Pusri	Fertilizer	Palembang, South Sumatra	Ammonia leak.
PT. Pakerin, PT. JS, PT. EA	Paper, shoes	East Java, Porong River	Damage to 1,300 ha of aquaculture ponds.
PT. IIU	Pulp	North Sumatra	Effluent ran into Asahan River in August 1988. Chlorine tank explosion in November 1993. Effluent holding pond broke open in March 1994, releasing effluent.
PT. IKP, PT. CP, PT. OPU, PT. SMP, PT. PJ	Paper, pulp, leather	West Java, Ciujung River	Damage to aquaculture ponds, paddies, etc. Environmental Impact Management Agency and NGOs contacted in 1993. Shrimp aquaculture company and villagers file lawsuit in 1995.
PT. Freeport	Copper and gold mining	Irian Jaya	In 1995 WALHI (Indonesian Environmental Forum) sued the government because it permitted the company to operate even though it was polluting.
PT. Adato Indonesia	Coal	South Kalimantan	Polluted Balangan River in March 1994.
PT. IMI	String, cloth, plastic bags	Central Java	Three affiliates of PT. IMI began operating in 1973. Water pollution. Sixty NGOs planned boycott of their products.
PT. MMII	Cans	Tangerang, West Java	Began operating June 1991. In June 1992 caused odors by burning waste paint, citizens protest. After repeat of problem in July, citizens break factory windows, etc.
PT. CSI	Monosodium glutamate	Pasuruan, East Java	In 1995 several thousand people damage the company's vehicles and buildings in the belief that it was causing the decline in shrimp production. Police arrest 63. Environmental Impact Management Agency announces that the company took adequate pollution control measures.

Source: Prepared with information from *TEMPO*, *Jakarta Post*, and other magazines and newspapers.

coordinates the program. Participating provinces increased to 11 in 1990, 13 in 1994, and 17 in 1995, and the program is limited to 15 industry types including caustic soda, electroplating, textiles, tapioca, paper/pulp, and monosodium glutamate. Each province designates its governor as the highest responsible official, and exacts pledges from the companies sited along the affected rivers that they will reduce their effluent amounts, then monitors to see if the companies actually do so. Because the program cannot begin with all the companies along a river at once, the number of companies is gradually increased. West Java, for example, started with 101 companies in 1989, then added 250 companies in 1990, 249 in 1991, and 200 each in 1992 and 1993. In 1994 the number of companies covered nationally had attained 1,900. In Jakarta and other places, company names are made public if they do poorly. The worst polluter, PT. TI, was forced to shut down for six months and install pollution control hardware. Water quality has improved in several rivers including the Musi River in South Sumatra. Still, it is only the two provinces of Jakarta and West Java that have been able to steadily increase the number of companies in the program and strengthen supervision over them. Up to 1994, in fact, 80% of the companies in the program were concentrated in these two provinces. by the end of 1993 the manufacturing industries had 140,000 business locations nationwide, with probably under 2% of them in the program. Some reasons that provinces can only gradually increase the number of companies, and must narrow the program to only certain rivers, are that the capacity of central and provincial governments to implement environmental efforts is considerably restricted by the number of technical personnel and funding limitations, and that regional differences are large.

Controls on hazardous wastes began in 1994 with the passage of a law regulating them. Just as with measures to clean up rivers, this law exacts promises from companies to reduce their waste amounts and then checks on compliance.

In 1995 Indonesia created air pollution standards covering motor vehicle exhaust and the four industries of cement, pulp, coal-fired thermal power, and steel. Using the same method as in remediating rivers, these standards involve a program to reduce emissions from point sources.

At this stage the government lacks sufficient personnel and financial capabilities to control pollution in all areas including water pollution, air pollution, and wastes, and to supervise all companies. For that reason it is dealing with pollution by gradually increasing the number of companies it properly supervises. Especially at the lower-tier local government level, such as regencies and municipalities, the capacities of environmental departments are limited. Indonesia needs a policy that will provide for training of environmental technical personnel and the enhancement of personnel and funding in local governments' environmental departments. It will also be necessary to obtain the cooperation of developed countries in such areas.

3. Forest Loss and Conservation

3-1 Precarious Balance: Forests and Economy in Java[7]

The mention of "Java" conjures up rural scenes presenting lowland expanses of farmland and terraced rice paddies reaching up to the sky, and this scenery has been preserved because the people have achieved a certain measure of livelihood through rural industries such as articles woven of grass, rattan, and bamboo, the manufacture of palm sugar and tofu, and batik dyeing. For example, a 1981 survey of nine agricultural villages whose main crop is rice demonstrated the importance of these industries.[8] In the villages as a whole, over 60% of income was non-agricultural. It is evident that Java's dense rural population is supported by a rural economy that assumes the development of the nonagrarian sector. We should therefore be aware that rural scenery dominated by rice paddies and teak forests has been maintained by the development of urban export industries and farming village industry. The history of forest utilization on Java resembles that in European countries, where forest loss has stopped since the time agricultural productivity rose with the advance of industrialization, allowing people to get by without clearing forests for farmland or cutting them for fuel.

Why does Java's history of forest utilization resemble that of Europe even though it was under colonial domination? The reason that Java avoided ecological catastrophe despite heavy plunder and the rapid loss of forested area was perhaps because, thanks to the fertile soil, people had a comparatively low dependence on the forests. Another significant factor is that during the colonial era government land was clearly demarcated, which built the foundation for management of national forests and led to the decline of swidden agriculture. Owing to these conditions, plus increased rice productivity and the development of labor-intensive industry under post-independence government-led industrialization, ecological balance has been somehow maintained despite the growing population. Thus Java's beautiful scenery actually exists in a precarious balance with the potential to be upset.

3-2 Forest Development in East Kalimantan

In outer Indonesia it is Irian Jaya and East Kalimantan that make a typical contrast to Java. A swidden ecosystem prevails in these regions with their lush forests and low population densities. Owing to development of the forests, the non-oil per capita GDP growth rate is high. Because the logging boom began in East Kalimantan in the early 1970s, it had a two-decade head start over Irian Jaya in forest development, making it a good subject for studying the process of deforestation.

East Kalimantan was saddled with a variety of problems owing to the rapid loss of forest in conjunction with forestry development starting in the 1970s. Considerable changes have occurred in the economic and natural environments of the Punan, who live by hunting and gathering in the hinterland, and the Dayak, who make their living with traditional swidden agriculture and by selling forest products.

• Industrialization and Transformation into a Market Economy

Let us begin by defining a concept to examine the development of forests. We shall define "transformation into a market economy," which progressed gradually over some centuries, as the process in which a non-market society takes on the character of a market society. This transformation was led by the petty merchants, who endeavored to derive commercial profit by taking advantage of ecosystem differences. Their activities incorporated the Dayak into the South-

---

[7] This section owes much to: Inoue, Makoto, "Forest Utilization and Economic Development in Indonesia," in M. Nagata, M. Inoue, and H. Oka, *Utilization and Regeneration of Forest Resources: The Logic of Economics and That of Nature*, Nobunkyo, 1994 (in Japanese).

[8] White, Benjamin and Gunawan Wiradi, "Agrarian and Nonagrarian Bases of Inequality in Nine Javanese Villages," in Gillan Hart, ed., *Agrarian Transformation, Local Process and the State in Southeast Asia*, Univ. of California, 1989.

east Asia trade network. Next we shall define "industrialization," the most modern phase of transformation into a market economy, as the dynamic growth process including the assumption of a central position by fixed capital goods, improved production efficiency through learning effect accumulation and greater division of labor, and the accumulation of external economic effect by means of integrated siting. This kind of industrialization is dominated by businesses, and the creation of credit by banks is a vital underlying element.

To be highly capable of adapting to industrialization, a society must have established market society forms for allocating and utilizing the production elements of labor and land. That is to say, creating the right to private land ownership and establishing the right to freely dispose and use labor must precede industrialization because free economic activity by merchants alone is not sufficient to adequately transact and allocate the production elements of land and labor.

• Industrialization: The Three Revolutions
1970s Logging Revolution

Dipterocarp trees in Borneo's lowlands grow as high as 70 meters, and this area is said to be the world's largest forest. Unlike the teak that Europeans had been after since early on, these forests were nearly untouched until recently because they had no commercial value. The trees in East Kalimantan have the best quality of those on Borneo, and logging was concentrated there in the 1970s.

Change in the way forests were developed came with the abandonment of *banjir kap*, a system of cutting and extracting timber that used human labor and the higher river water levels during the rainy season, and the adoption of capital-intensive methods that involve the use of chain saws for cutting, heavy machinery for extraction, and trucks and boats for transport. Most of the workers were Javanese who had immigrated or were there just to earn money. This system exponentially increased East Kalimantan's log production, raising its log exports from 300,000 cubic meters in 1968 to more than 7 million cubic meters in 1978. Throughout the 1970s, East Kalimantan accounted for one-third to one-half of Indonesia's log production.

Meanwhile, intensively logged land turned into extensive swidden or pepper fields. Because these fields were abandoned as their fertility declined, the agricultural frontier kept moving farther into the forest interior in search of new logged-over land, resulting in the continual shrinking of forested area and the appearance of barren grasslands like *alang-alang*. Further, the extensive forest fires of 1983 burned over much of the logged area. In only about 20 years the forested views of East Kalimantan completely changed except for the hinterlands.

1980s Plywood Revolution

In 1970 immediately after the logging boom began, the Suharto government issued an order requiring all companies with logging concessions (HPH) to establish forest product processing plants, such as for plywood, within three to 10 years after they start logging. Although HPH companies thus started building plywood plants in 1973, at first these were almost all producing for the domestic market. But a 1980 change in forest policy obligated HPH companies to supply logs for domestic use. Restrictions on log exports gradually tightened, ending in a ban in 1986. Indonesia's development policy had switched from emphasis on the export of primary products like oil, natural gas, and logs, to one emphasizing exports of industrial products. An effect of this was the rapid development of the plywood industry into an export industry after 1979, with plywood exports of about 120,000 cubic meters growing 30-fold in six years to over 3.5 million in 1985 and to 9.6 million in 1993 (3.7 million cubic meters went to Japan). In 1985 plywood outstripped traditional export industries such as coffee, tin, processed rubber products, and shrimp, coming out in third place behind oil and natural gas. In monetary terms the export ranking in 1993 was textiles, oil, plywood, and natural gas. Growth of the plywood industry contributed to growth in adhesive manufacturing and other related industries, making plywood's unmistakable contribution to the national economy an employment increase. A mere 742 plywood workers in 1974 ballooned to 14,800 in 1979, 118,000 in 1984, and 445,600 in 1993, making the plywood industry one of those with the biggest employment capacity among large and medium-sized indistries. East Kalimantan is of central importance in the plywood industry. There are so many plywood factories along the Mahakam River on the outskirts of Samarinda that the area is called the "world's plywood mall." In 1993, 27 of Indonesia's 121 plywood factories were in East Kalimantan, where the large number of people and the high per capital GDP attest to the plywood industry's importance to the economy.

1990s Tree Plantation Revolution

In the 1990s Indonesia's government energetically promoted "industrial tree plantations" (HTI), which signifies the planting of production forest land by business concerns that have obtained an "industrial tree plantation concession" (HPHTI). Its purposes are to strengthen the country's wood industry, while promoting environmental conservation by planting trees in deforested areas. HTIs are areas used to produce chips for pulp, and areas for other purposes, but in either type of area industrial plantations (including clearcutting) are meant only for low-productivity areas of under 25 cubic meters per ha. Organizations with HTIs can be divided into four categories based on their characteristics: Provincial forestry bureaus, the national forest products company (INHUTANI), government-private sector joint ventures, and private businesses/cooperatives. Many HTI organizations are joint ventures because they get preferential treatment such as 14% of costs provided by the government, and they need only 21% owned capital.

However, many people have already settled on lands to be reforested, where they practice swidden agriculture. Some of these people are the indigenous Dayak peoples. Companies have varying ways of taking over lands to be planted. If the settlers are Dayak, for example, even though they may not be forced to give up land actually under cultivation, if their gardens are surrounded with planted woodland, they will be forced to move out in a few years when the fertility of their farmland declines. By that time fallow forest lands will have been logged over and replanted with fast-growing species, leaving the settlers with little space for swiddens. Losing their land in this manner forced people into considerable lifestyle changes. Of course, lumber companies pay them compensation when their perennial crops have been planted on land to be forested; allow the people to intercrop food crops among the planted trees for one or two years in return for managing the trees; lend guidance in planting trees needed in their livelihoods; and provide other assistance. Basically, however, the only two choices that settlers in industrial tree plantation areas have are to either become forestry workers or get out.

Forest policy is shifting from the natural forest logging phase to the plantation phase. While this shows the failure of sustainable forestry that consists mainly in selective cutting, it is also the unfolding of an orthodox forestry policy that aims to create productive forests by planting trees on degraded land. Currently, however, the friction between forest inhabitants and forest policy, which was not very evident during the natural forest logging phase, is becoming evident over land expropriation for industrial tree plantations.

• **Forest Development as "Incomplete Industrialization"**

As we have seen, the element common to the above three industrialization processes is that the Dayak and other forest-dwelling native peoples have not benefitted much from development, or have been adversely affected. Their habitations have been incorporated into national forests, and companies with concessions have plundered their means of subsistence. Land transaction and allocation, which were to be accomplished for industrialization, were therefore to an extent achieved by force. On the other hand, most of the loggers and afforestation workers are Javanese, with a very small proportion of Dayak. The task of labor transaction and allocation was thus achieved by hiring Java's excess labor.

In view of this situation, it was not the people of Kalimantan, but the companies and Javanese backed by national policy, who had the capacity for adapting to industrialization. East Kalimantan is the scene of "incomplete industrialization," i.e., industrialization that is proceeding despite the affected societies' lack of adaptive capacity.

3-3 Coexistence of People and Forests

• **Two Complementary Policy Concepts**

As "incomplete industrialization" proceeded rapidly in East Kalimantan, forests disappeared and social problems arose over land. This situation leads to the logical conclusion that forest policy henceforth should take one of two directions: Either try to achieve "complete industrialization," or take a different tack, that of rebuilding a social system based on the commons.

Grounds for the "complete industrialization" strategy are that the Dayak themselves wish to benefit from modern civilization, and that as Kalimantan has a long history of transformation into a market economy, the conditions for industrialization are readily found. If this strategy is adopted, it would be necessary to set up a system allowing the Dayak to sell their land and labor as they like, which would in turn require education and private land ownership. Under this system forestry policy, including industrial afforestation and its potential problems, would move in tandem with policy guaranteeing ownership of residential and farming land, and educational opportunity. Intercropping in the industrial tree plantation areas would be allowed only transitionally.

Grounds for the "commons" strategy assume that although people want a market economy, they do not necessarily desire industrialization. This would necessitate giving full play to the Dayak system of using the forests while keeping intact the Dayak-merchant trading network that has existed since before modern times, and would rebuild autonomy and social relationships that are not subject to official control. Social forestry — forest-related activities with local people's participation meant to stabilize and improve their livelihoods and welfare — would play a central role. Specifically this system would involve, for example, leaving the management of protected areas to local people, who have long had a sustainable relationship with the forests, and allowing "communal forests," in which inhabitants continue customary forest utilization to complement the complete industrialization policy.

• **Zoning Based on Current Land Use**

Implementing policy concepts like those above would consist basically in clearly demarcating forest zones and determining ways to manage them appropriately in accordance with their purposes while taking the livelihoods of local people into account.

Indonesia has classified its forests on the national and provincial levels into nature reserves, protection forests, limited production forests, production forests, and conversion forests. This zoning scheme has various problems, the most serious being that current land use is not taken into account at all when drawing zone lines on the map. As inhabitants are thus left out of the equation, it is difficult to give them a role in managing forests whether complete industrialization or the commons system is chosen. A zoning system should take into account the inhabitants and their use of the land (Table 4). In some nature reserves it will be necessary to leave management to the inhabitants, which would involve restoring the commons. UN data on protected areas show that with 180,000 square km protected, Indonesia ranks third after Brazil and Venezuela among tropical forest countries, while in number of designated places it ranks first with 169.[9] Owing to Indonesia's importance in protecting world biodiversity, there is concern about effective methods of

---

9   IUCN, *United Nations List of National Parks and Protected Areas*, 1990.

Table 4 Basic Policy on Forest Use Categories (Proposed)

Forest category		Purpose	Vegetation	Inhabitants	Forestry activities
Nature reserves	Strictly protected areas	Conserving biodiversity.	Primary forest	None.	None.
	Areas for hunter-gatherers	Guarantee land to hunter-gatherers. / Production of non-wood forest products. / Conserving biodiversity.		Hunter-gatherers.	
	Areas for traditional swidden farmers	Guarantee land to traditional swidden farmers / Production of non-wood forest products. / Conserving biodiversity.		Traditional cyclic swidden farmers.	
Protection forests		Erosion control, watershed protection, etc. / Production of non-wood forest products.	grassland	May be expelled in certain circumstances.	Environmental afforestation (when necessary).
	Communal forests	Guarantee customary forest use to indigenous peoples.	Secondary forest	Semi-traditional swidden farmers.	None.
Production forests		Wood production. / Production of non-wood forest products.		Forestry workers.	Selective cutting, industrial tree plantations. / Enrichment planting. / Social forestry.
Conversion forests		Conversion to farmland and residential lots.	grassland	Non-traditional swidden farmers.	Social forestry (especially farm forestry).

Notes: 1) Except in strictly protected areas, hunting is allowed to the extent needed for the livelihood of indigenous peoples.
2) Secondary forests include logged-over forests and former swiddens.
3) Social forestry in production forests may assume many forms, such as fuelwood plantations, taungya-type agroforestry, etc.
4) Communal forests exist in spots throughout production forests.

Source: Prepared by Inoue on the basis of field work in Indonesia.

managing its protected areas. While the strategy basically applied in production forests might be complete industrialization that allows private land ownership, it might also be necessary to adopt the commons idea, such as by allowing some communal forests. Doing so would give inhabitants an incentive to protect the forests, thereby providing the government with a low-cost forest management method. This would of course assume detailed economic and social studies. In conversion forest areas, it would be important to promote agroforestry to ensure that farmland is used sustainably.

• The Two Strategies in National Forest Policy

Finally let's examine the possibility of these strategies becoming reality. The 1960 Basic Land Law recognizes customary communal disposal rights and establishes ownership based on customary law. This makes both strategies possible as long as they do not oppose the interests of the citizens or the state. Under the 1967 Basic Forestry Law, customary forest use is possible as long as logging companies allow it. So although it is impossible to place a whole region under the dominion of its inhabitants, it is possible to secure a portion of a company's commercial forestry area as a communal forest. Such an arrangement would be the partial restoration of the commons under complete industrialization.

But the Basic Forestry Law makes it impossible for a group of inhabitants occupying a certain parcel of land to obtain a customary communal disposal right, which rules out a policy that leaves the management of protected areas to indigenous peoples. If the government strictly applies the Basic Forestry Law and promotes the confirmation of private land ownership as set forth in the Basic Land Law, then the commons strategy will become quite unrealizable. In fact, land registration is underway with World Bank financing, which means one can only watch to see that complete industrialization is properly implemented.

Nevertheless, there are attempts to facilitate social forestry and local participation as part of national forest policy, which would probably necessitate explicitly allowing customary communal forests by changing the Basic Forestry Law provisions stating that tribal forests, local forests, local government forests, and the like do not exist. Having done so, it would be crucial that the Dayak and other forest dwellers are able to set up a collective forest management system with the help of the government and NGOs, which would only then make it possible to at least partially incorporate the commons strategy.

4. The Importance of NGOs in Environmental Policy

4-1 Citizens' Movements in Pollution Disputes

Indonesia has frequent pollution disputes, with some going into arbitration and the courts, and others escalating into violence. This section will examine: how, when citizens call attention to pollution problems, this can lead to pollution control action by businesses; how pressure is brought to bear against such citizens; and the roles played by local governments, the media, and NGOs.[10] We'll begin with a representative case.

• Pollution of the Tapak River, Semarang City

A lime citrate manufacturer, SDC, began trial production in 1976 on the outskirts of Semarang City in Central Java, and in 1978 started commercial production. Because the company ran its effluent into the Tapak River, which is only two or three meters wide, downstream rice paddies and aquaculture ponds were contaminated, causing decreased yields and fish mortality. Citizens complained to the mayor repeatedly beginning in February 1977 but owing to an inadequate response the company and citizens held a meeting at which the company agreed to abate its pollution, and to supply the citizens with water until pollution control steps were implemented. However, as the agreement was not documented, and because implementation was not adequately verified, SDC did not do enough.

Subsequently seven other factories making a variety of products were sited near SDC, and water quality worsened. In 1985 poor water quality killed fish and shrimp in aquaculture ponds. Citizens complained to the city, but owing to the inadequate response they formed an organization, appealing and applying persuasion to the provincial and central governments, and to NGOs. The extent of damage became evident through surveys by WALHI and Semarang City authorities in 1986, and Diponegoro University in 1989. Because citizen requests to the mayor, the minister of industry, and the vice-president led nowhere, 15 environmental organizations announced a boycott plan in April 1991 to pressure the government and companies, exacting payments of 225 million rupiah in compensation and 185 million rupiah in pollution measures from eight companies in August 1991.

• Circumstances Affecting Citizens' Movements

Citizens' movements and the political and social circumstances that affect them can be summarized as follows.

Citizens begin by complaining to factories about their pollution, and in the absence of results they gradually take their complaints up the administrative ladder beginning with neighborhood associations. Because local political entities in many areas have no specialists to deal with environmental problems, and because investigations, even if launched, rarely result in concrete action, citizens take their complaints to regional governments and the House (parliament), and appeal to the media and NGOs.

Going directly to court is difficult because citizens usually lack sufficient legal knowledge, and because the Environmental Management Act stipulates that before going to court the disputants must attempt a compromise settlement.[11] Under the Act's provisions, the government establishes an arbitration team consisting of the victims and polluters (or their representatives), and government employees. If the team cannot agree on what damage has occurred and its monetary amount, the matter is to be settled in court. Even then, however, the citizens may not be able to prove their case, and even if they win, they might not be compensated. For example, in 1994 people who would be displaced by the Kedung Ombo Dam won a court victory for monetary compensation, but a Supreme Court judge ordered a suspension.

The media are rather well-disposed toward citizens, and many pollution disputes make the news. Nevertheless, the media are subject to government pressure, with three weekly magazines banned in 1994. In some instances citizens who appealed to the media or otherwise indicated their indignation found themselves summoned and warned by the military or other authorities.

As regional leaders are appointed by the central government — provincial governors by the president, regency governors and the mayors of muncipalities by the minister of Home Affairs — they seldom turn an attentive ear to what the people are saying. Although local assemblies and the parliament occasionally deal with pollution problems, they do not have that much influence. Both local assembly and parliament members are chosen in proportional representation elections that allow participation only by three factions in whose operation the government intervenes. No other political parties may exist. Even elected representatives can be relieved of their posts if deemed unqualified.[12]

In some instances violence arises if the citizens' complaints do not lead to a resolution. At Tangerang in West Java in July 1992 a malodorous tin can factory was attacked by over 250 local citizens.[13] And in November 1995 in East Java violence was triggered by protests from fisherfolk who claimed that factory effluent had polluted shrimp aquaculture ponds.[14]

• Expectations for NGOs

Indonesia's NGOs play a major role in keeping violence from occurring and solving environmental problems. An example would be the Indonesian Legal Aid Foundation

---

[10] One researcher distinguishes between "community-based movements" by people whose interests are directly involved, and "citizens' movements" by highly educated, conscientious citizens. By "NGO" here we mean organizations with a strong "citizens' movement" character.

[11] Under a 1997 revision of the Environmental Management Act, a dispute can be taken to court even if arbitration efforts have not been made.

[12] The resignation of President Suharto in 1998 engendered major changes in the political system, such as the freedom to form political parties.

[13] *Kompas*, July 31, 1992.

[14] *Jakarta Post*, November 22 and 23, 1995.

(YLBHI), whose purpose is to defend human rights, and whose lawyers provide consultation on environmental problems. If there is a citizen request, they talk with the environmental departments of the central and provincial governments, also participating in arbitration teams and working to safeguard the citizens' rights. In the Semarang City case described above, WALHI and YLBHI ran a boycott campaign to pressure the authorities into an arbitration agreement. But there is also government pressure against NGO activities. High government officials have publicly voiced negative views about boycotts, and the government also applies pressure against the anti-nuclear power campaign by WALHI and other organizations by not accepting meeting notifications, thus forcing meeting cancellation.

In order to solve environmental problems it is crucial that citizens who sustain pollution damage be allowed to air their grievances. Further, at a time when authorities responsible for dealing with pollution lack personnel and funding, it is important to beef up surveillance of pollution sources with emphasis on those that elicit citizen complaints. Indonesia must provide for institutional improvements, such as establishing procedures to be followed until arbitration teams are formed, or opening regional government offices to handle pollution disputes. Also needed are institutional reforms that will more readily allow citizens to make themselves heard and have their views incorporated into policy, with such reforms including those on local autonomy and judicial systems, and establishing freedom of the press.

4-2 Citizens' Campaigns in Forest Conservation

Off the west coast of Sumatra lie the Mentawai Islands, whose biggest island is Siberut. As over 60% of Siberut's flora and fauna are endemic species, the island is in the spotlight for its importance to conserving biodiversity. But since 1960 the government's resettlement policy has increased the island's population and encouraged rice agriculture. Furthermore, logging under concessions, the development of oil palm plantations, transmigration operations, and the like are contributing to environmental degradation, thereby threatening the people's livelihoods. Following is a brief overview of the NGO role in establishing a national park.

• An International Campaign

1990 saw the formation of SOS Siberut, an international organization based in Britain. SOS Siberut mounted a campaign to protect Siberut with the cooperation of SKEPHI (Indonesia Forest Conservation Network) in Jakarta and PMS (Siberut Residents Association) on the island itself. As Siberut was also designated a protected area under UNESCO's Man and the Biosphere (MAB) program, the campaign enjoyed strong international support. Campaigners conducted a letter-writing campaign, and also lobbied the British government, UN, and other bodies, as well as WWF and UNESCO, which had been involved with the Siberut issue since the early 1980s. In November 1991 the campaign issued a newsletter worldwide appealing for urgent action, which resulted in a huge volume of written demands and protests sent by the people of many countries to Indonesian government ministries.

Table 5 Indonesian Environmental Policy Timeline

Year	Policy and Other Changes
1967	Basic Forestry Law, Foreign Investment Law.
1968	Domestic Investment Law.
1970	Foreign Investment Law revised to provide for selective introduction of foreign capital.
1974	Forest development concessions limited to domestic capital.
1977	Proclamation by mayor on water quality in Jakarta's rivers.
1978	Appointment of State Minister for Development Supervision and Environment; provincial governors' decisions on industrial effluent in East Java and Central Java provinces.
1980	Law on reforestation fund to obligate concession holders to contribute US$4 per cubic meter.
1981	Phaseout of log exports; policy to foster plywood industry.
1982	Environmental Management Act.
1983	Appointment of State Minister for Population and Environment.
1985	Total ban on log exports.
1986	Regulation on Environmental Impact Analysis.
1988	Decree by population and environment minister on environmental quality standards; no more permits for new plywood factories.
1989	PROKASIH (Clean River Program) initiated; reforestation fund increased (US$7 per cubic meter); legal sanctions enacted on violations of forest development provisions.
1990	Environmental Impact Management Agency established; reforestation fund increased (US$10 per cubic meter); law to facilitate industrial tree plantations (HTI).
1991	Ministerial Decree on Waste Water Quality Standard for Activities Already in Operation.
1992	Repeal of total ban on log exports and institution of high duty on their exports (US$500-4,800 per cubic meter).
1993	Appointment of state minister for the environment.

• Domestic NGO Activities and the Government Response

Since the beginning of that international campaign, the Siberut Island issue has been a very sensitive one to the government. In fact, a workshop to be held in West Sumatra was forced into cancellation by the Directorate General of Nature Protection, Ministry of Forestry. In December 1991 SKEPHI, with the participation of the media, the government, and NGOs, scheduled a seminar in Jakarta in order to provide the people of Siberut with an opportunity to speak their minds, but two days before the scheduled date a warning forced it into cancellation. PMS and SKEPHI then talked to the Ministry of Population and Environment, which, although politically weaker than the Ministry of Forestry, had indicated an understanding of the Siberut issue. This catalyzed a meeting of SKEPHI, PMS, and representatives of several ministries, with the minister of population and environment playing the leading role. Although SOS Siberut and the media were not allowed to participate, the meeting achieved major progress including decisions to send a special government fact-finding mission to Siberut, and to hold inter-ministry meetings on the matter.

These points summarize the response of Indonesia's government: (1) The Ministry of Population and Environment has consistently taken the side of nature protection and native rights. (2) However, it has been powerless against the Ministry of Forestry because of its own weak political strength. (3) The Ministry of Forestry's overriding priority is wood production. (4) Yet, there is also a Directorate General of Nature Protection (PHPA), which has created protected areas under pressure from international public opinion and other factors. (5) Nevertheless, the Ministry of Forestry tends to avoid the issue of native rights.

• Creation of a National Park

On March 31, 1992 President Suharto issued an order under which logging on Siberut Island would conclude upon the expiration of the present forestry concession. There was also a decision to adopt a new approach: Instead of depending mainly on patrolling, Indonesia would try to conserve its biodiversity by combining the management of protected areas with community socioeconomic development achieved through local participation. Plans were made to translate this policy into concrete action as a biodiversity conservation project under the credit agreement signed with the Asian Development Bank on December 21, 1992. According to the provisions, within six months after the agreement went into effect, the Ministry of Forestry was to officially designate as a national park the Siberut Island wildlife sanctuary and the adjoining logged-over forest that was designated a production forest, and on August 10, 1993 the Minister of Forestry indeed designated 190,500 ha (43% of the island) as Siberut National Park. Thus, NGO support for local indigenous peoples, with the backing of the MAB program and the support of international public opinion, demonstrated that NGOs are capable of considerable achievements.

5. Summation

Generally it is the central and regional governments that are the primary actors in environmental policy. Such being the case, while the government determines the system for policy objectives and means, it is supposed to incorporate citizen preferences by way of a policy determination mechanism. But because governments often represent mainly the interests of certain groups, one cannot automatically assume that a government "democratic" in form will choose policy objectives and means that accurately reflect the will of the people. Accordingly, one hopes not only that environmental policy has a public sector (the government), but also that NGOs and businesses too will play important roles as primary actors.

Yet, in a parliamentary democracy it is legislation by the parliament that is the proper means of placing environmental policy under control of the citizens. However, it is unrealistic to restrict everything with laws, and thus in view of specialization, suitability to the situation, and flexibility, one cannot deny the need for administrative discretion. Vital here is democratic control over administrative planning, which is where administrative discretion manifests itself. Of cardinal importance to that is freedom of information, which provides citizens with the basic information they need to make judgments, and local citizen participation, which guarantees them the opportunity to be involved.

In any event, local participation will henceforth be the keyword of environmental policy, although hardly any country in the world guarantees this institutionally, Indonesia being no exception. This means that movements by local people will perforce play a bigger role in environmental policy. Such movements arise spontaneously in response to harm and oppression committed in the name of "the public interest," and constitute an antithesis to domination by public power and business. When policymaking processes are closed, these movements generally tend to be little more than campaigns to censure or resist, and it is difficult to build their capacity to offer alternatives that are constructive and feasible. Indonesia too needs to create an enabling political environment in which community-based area resident movements (by people living in the affected area), and the alliance-like citizen movements (by national and international NGOs) that support them, can commit themselves to constructive work on environmental problems. To facilitate this process of working toward self-initiated and sustainable development it will be essential to have not only cooperation by Japan's government, which has a highly visible presence through official development assistance, but also international and Japanese citizen support for Indonesian NGOs.

(INOUE Makoto, KOJIMA Michikazu)

Essay 1   Change in the Swidden System

The chief habitation of the Kenyah Dayak, a swidden farming people of Borneo, is the Apo Kayan region in the hinterland. In the early 1950s they began migrating to the watersheds of the Mahakam and other rivers, and now have villages throughout East Kalimantan.

They classify the vegetation of former swiddens into several categories and name it depending on its ecological succession stage. Their livelihood depends on recurrent swidden farming in which reuse of land waits until tree trunks are at least as big as a person's thigh, which normally takes 10-odd years. Important here is that the determinant of this cycle is not how many years have passed, but the recovery phase of vegetation in former swiddens. This system is ecologically sound because even if site conditions vary, it assures there is a certain amount of nutrients available.

But their system is undergoing rapid change. Some people in downstream villages now reuse sites that still have grassy undergrowth. Also, the farther downstream a village lies, the more it is integrated into the market economy, and the more changes there are in customary land holding systems, mutual help arrangements in daily life, and the organization of labor for swidden agriculture. Originally the Kenyah Dayak's subsistence depended completely on the stability of forest resources, thus making it necessary to avoid taking too much from nature, which would invite a reduction or exhaustion of resources. These changes therefore constitute a process that weakens their role as forest conservationists.[15]

(INOUE Makoto)

---

[15]   Based on a field study by Inoue.

Essay 2   The Process of Deforestation

Indonesia's government classifies forested land into "nature reserves," "protection forests," "production forests," and "conversion forests." Companies that obtain concessions, which are mainly on production forests, are obligated to observe certain restrictions, such as waiting to log again only after 35 years, but concessions are normally 20 years. Although extensions are possible, companies have a stronger motive to make the biggest possible profit in a limited time than to use resources sustainably by thinking ahead 35 years. In fact it was often the case that companies brought in heavy machinery and recklessly took out the trees with high commercial value. While it is the government's job to prevent this, it is impossible for the limited number of workers to monitor all the remote logging sites. And even if companies follow the rules by leaving trees that promise to be valuable someday, it is very possible they will be logged illegally, another factor heavily influencing logger behavior.

In 1980 about 100 loggers had concessions in East Kalimantan, but there were also 50 to 60 logging teams operating without government permission, which apparently cut the remainder of forests that had been selectively logged by concession holders mainly along the road joining the provincial capital of Samarinda with the oil town of Balikpapan. The decisive push toward deforestation came from settlers who used the roads through logged areas to move into the forests, where they practiced non-traditional swidden farming. On the lower reaches of rivers in East Kalimantan settlers created pepper gardens that are maintained by thorough weeding that exposes the ground to rain and direct sunlight, resulting in the nearly complete loss of topsoil. When people stop maintaining the gardens owing to production declines or a fall in the market price, the land is taken over by *alang-alang* grass and turned into prairie. Typical throughout Southeast Asia is this deforestation process, which begins with commercial logging and ends with non-traditional swidden farming.

(INOUE Makoto)

Chapter 6
People's Republic of China

1. Introduction

The People's Republic of China has the world's largest population and the third-largest land area, and is one of the biggest countries in terms of ecological scale. Since the 1980s it has been one of the few countries to continue long-term rapid economic growth. These facts are reflected in much research on China's environmental problems and initiatives. Observations made thus far on the characteristics of China's environmental problems can be summarized as follows.

(1) China's environmental problems are already of crisis proportions.

(2) With its population/resource imbalance, continental topography having many mountainous areas and encouraging long pollution residence time, vegetation that recovers only with difficulty once disrupted, and other such factors, China's natural conditions present far greater disadvantages than other countries and regions.

(3) China's civilization of several thousand years, invasions by European powers and Japan, civil wars, the trial-and-error process of building socialism, and other factors have imposed a huge burden on the land.

(4) The financial combines and other powers (especially Japan) before and during the war, and the Communist Party after the war, structured an economy involving mainly heavy industry and coal consumption, thereby creating a huge environmental burden, and that economy continues to operate full tilt even under the policy of reform and opening the country.

(5) From the end of this century and through the first half of the next, China will plunge into the mass-consumption, mass-disposal society, progressing from urban to rural areas.

In view of these characteristics, it is perhaps inevitable that economic development and environmental damage go hand in hand. China is as committed to development as any other country; the process has been set in motion, and the destination is unknown. Such being the case, the only course of action is to cooperate with many countries and people in making course corrections, however minor they may be, which will benefit not only China, but all of Asia and the world.

This chapter will begin by using specific examples in a survey of China's regional environmental problems, township and village enterprises (TVEs), and the Chinese government's new initiatives, which have not had adequate treatment thus far. It will then quantitatively evaluate China's impacts on the global environment and examine a scenario for building a regional environmental security system in East Asia.

2. Overview of China's Environmental Problems

2-1 Regional Survey

• Northeast: Pollution by the Heavy and Chemical Industries

Concentrated in this region is pollution by the heavy and chemical industries that reminds one of Japan's worst era. Compared with China as a whole, Liaoning Province in particular has a large industrial waste volume and high emissions of other substances. Also found in Liaoning is the city of Benxi, which at one time was apparently blocked from the view of satellites by severe air pollution. Also occurring in this region in the early 1970s were the pollution of Dalian Bay and mercury contamination in the Songhua River. In considering the history of the industrial pollution that is a major cause of environmental problems in this region, one cannot discount the effects of prewar Japanese policy on the continent, especially industrial policy in Manchuria. Liaoning Province's industrial cities and main pollution sources of Shenyang, Fushun, Anshan, Dalian, and others were all forced into industrialization by Japan. The factory that contaminated the Songhua River with mercury did so by the same mechanism and with the same result as at Minamata in Japan.

Japan's military activities also left their mark on China. Beginning in 1991, Japan and China investigated the chemical weapons that had been abandoned in China, which may be a cause of air, water, and soil contamination. As a result, Japan has assumed full responsibility for their removal. The largest cleanup job consists of an estimated 700,000 chemical shells that China's government gathered immediately after the war and buried near Dunhua City in Jilin Province. According to the Chinese government, Japan's military left behind about 2 million poison gas shells and as much as 1 million tons of chemicals, 90% of which are concentrated in the Northeast. Three fact-finding missions dispatched by Japan's government in 1995 discovered abandoned chemical weapons quite possibly made by Japan's military, also in Jilin, Harbin, Shenyang, Hangzhou, Nanjing, and other places. While Japan tends to see China as the chief source of transboundary pollution, China must cope with an unwanted legacy of the past that creates environmental problems in the present.

• North and Northwest: Coal Pollution and Desertification

Coal is so widely distributed in China that about 90% of its cities can obtain it nearby to some extent, but there is more in the north, with Shanxi Province producing about 30% of China's coal, and with Xinjiang and Inner Mongolia having by far most of the coal reserves. Thus, while environmental problems associated with coal production are found nationwide, they are especially typical in the North and Northwest. Shanxi Province in particular, known for its nationally high coal production, has a great deal of environmental damage owing to coal mining and transport. Damage to aquifers by coal mining not only injures workers, but also aggravates northern China's water shortage. There are also reports of health and crop damage by dust from coal carriers along Shanxi's road and railway artery from Yuncheng to Datong.

In recent years there are large fires in mine shafts and coal fields. In 1996 there were many shaft fires in Ningxia, known for its high-quality coal, and in Xinjiang's coal fields, which are highly volatile. According to reports, about 6,000 lives are lost yearly in coal mining accidents throughout China.

Desertification is worst in Xinjiang and Inner Mongolia, caused by the opening of farmland in ecologically fragile areas. Salinization of grasslands, fires, and other factors are spurring on the destruction of land resources. In the Kuitun area of the Xinjiang Uighur Autonomous Region and in part of Inner Mongolia, drinking water from deep wells caused many cases of fluorine and arsenic poisoning. At one time there were as many as 2,000 cases of arsenic poisoning, and even some people poisoned with both arsenic and fluorine. Reports from Inner Mongolia say there are several million victims of fluorosis and 1,600 arsenic poisoning victims, as well as many people with both, and many more latent cases.

• **West and Southern Border Region: Transboundary Problems**

It is the Himalayas and the Mekong River where China most directly faces transboundary environmental problems, but perhaps because of the Tibet issue, there is little discussion of Himalayan problems as they involve China. But countries throughout the Mekong River watershed are growing keen on development, and there is word that China has dam building plans. The region from Yunnan Province to the Guangxi Zhuang Autonomous Region is rich in wild species, especially wild animals, but due to common borders with Myanmar, Laos, and Vietnam, and the close proximity of Thailand, there is much poaching of protected and rare animals, as well as smuggling and illegal sales in border zones. To combat this problem, Chinese authorities clamped down on violators, imposing the death sentence on serious smugglers in November 1995.

Xinjiang and Tibet are in the environmental spotlight because of their nuclear development facilities. One of the Dalai Lama's demands to the Chinese government is the removal of nuclear waste facilities from the Tibet Autonomous Region. All of China's atmospheric and underground nuclear tests were conducted in Xinjiang, and in the 1980s it was where China provided the former West Germany with a nuclear waste disposal site. Behind this location of nuclear sites is the low population of China's western half — only one-tenth that of its eastern half — and the many minorities.

• **Inland: Stagnant Pollution and Agricultural Impacts**

Chongqing City in Sichuan Province is often named as a place with serious acid rain, and in fact its SO_2 concentration and respiratory illness incidences are worse than those in Yokkaichi, Japan were at their worst. Many places in China's interior such as Chongqing, and Guiyang City in Guizhou Province, have grave air pollution in comparison to their levels of economic development and industrialization. Common to all these places are the high sulfur content of their coal, and location in basins that tend to trap pollutants, but which are also the main agricultural producing areas in each province. Conditions for agricultural production are poor in Shanxi and good in Sichuan, but in both provinces industrial pollution has highly adverse effects on agriculture. Pollution sources are the TVEs found throughout rural areas, and the cities in the basins. In Guizhou and Hunan provinces there are reports of poisoning due to household use of coal containing arsenic.

• **Central China: Huai River Method and Three Gorges Dam**

The Huai River is between the Yellow and Chang Jiang (Yangtze) rivers, and its watershed is mainly in the four provinces of Henan, Anhui, Shandong, and Jiangsu. It has for some time been heavily polluted by factories including paper and pulp, and in 1994 and 1995 there were some especially serious accidents. In order to deal with pollution in the Huai, the State Environmental Protection Commission (SEPC) decided to close all small paper mills under 5,000 tons yearly production by July 1996. The four watershed provinces organized teams to phase out the mills, and the pulp manufacturing facilities in all 999 plants subject to the shutdown were closed by the deadline. Some people voiced concerns and doubts about the economic repercussions of forcing plants to close, while others worried that it would not long be effective. However, the SEPC has high praise for this Huai River method, and has indicated its intention to use it in other watersheds and economic sectors.

Meanwhile, work on the world's largest dam is in progress at Three Gorges midway up the Chang Jiang (Yangtze), China's largest river. Three Gorges is the place where the river descends from the highlands to the plains, and it has sites that are valuable in terms of both natural and Chinese history. It has also been long regarded as the place most suited to development in terms of electric power demand and hydropower potential. Current arguments for the dam include flood control, hydropower, water transportation, aquaculture, and water resource supply, but as the plan was fleshed out, domestic arguments against the dam appeared. These included an excessively large investment for the Chinese economy, the relocation of a million people, and disadvantages to national defence, as well as many environment-related rebuttals such as siltation, downstream soil deterioration, submergence of China's natural and cultural legacy, scenery changes, and ecosystem impacts including the extinction of the Chinese river dolphin and other species. In 1989 a book compiling these opposing arguments[1] was published, but then banned, which strengthened the resolve of the intellectuals and helped lay the spiritual foundation for the democratization movement that was crushed at Tiananmen. Even the 1992 resolution by the National People's Congress that made the final decision to build the dam included an unprecedented 33% votes in opposition or abstaining.

There has been heavy criticism from abroad as well. In 1988 the World Bank warned that the project is economi-

[1] See: Dai Qing, *Yangtze! Yangtze!*, Earthscan Publications, 1994.

cally impossible, and even after construction began the U.S. entirely suspended its scheduled technical and financing assistance, and forwent participation in international bidding for generating equipment. Many people believe that behind China's construction of this dam despite such opposition is its acute need for electricity and a sense of crisis over the frequent floods in recent years, as well as the importance of controlling the river as a face-saving measure.

• South China and Coastal Cities: Modern-Day Pollution

Special economic zones, established in 1980 to produce exports, became major focal points for foreign investment, and their success led to the establishment of economic and technological development zones around the country, mainly near coastal cities. These zones are less prone to preexisting types of pollution because they bring in foreign-capital firms with higher technologies than Chinese companies, and Chinese firms with advanced technologies. But it is very likely that they will bring about the same problems as industrialized countries have, such as high-tech pollution, and the danger is even greater because, depending on the investment origin, efforts to address safety and environmental concerns tend to be neglected.

Meanwhile, in 1996 Beijing and Shanghai discovered the presence of "foreign wastes," or wastes smuggled in from other countries, showing that China has the typical developing-country problem of imported wastes.

2-2 Pollution by TVEs

While the economic development of rural areas is definitely high on China's agenda, the trend toward an increasing urban-rural cash income disparity continues unabated, and the excess rural work force is predicted to be 150 million in 2000. TVEs, which are meant to increase cash income and absorb excess labor, are now an essential part of not only the rural economy, but the whole Chinese economy. Yet these rapidly growing TVEs also cause serious pollution, and for that reason China's government conducted a detailed national survey of TVE-related pollution problems from 1989 to 1992. This section will discuss the report by this survey and present as a case study an independent survey by one of the authors.

• TVE Survey Overview

(1) Survey Scope

The survey covered 571,200 companies throughout China, excluding the Tibet Autonomous Region, accounting for 7.8% of China's TVEs, and over 186 billion yuan or 35.5% of total TVE production at that time. Of these surveyed companies, 364,400 were "polluting industries," or those such as chemicals or papermaking that are highly polluting.

(2) Results

Water Pollution   The surveyed TVEs had an annual total of 1.8 billion tons of effluent, of which 280 million

tons (15.6%) satisfied the State Environmental Protection Agency's emission standards. Polluting industry companies had 1.37 billion tons of effluent, of which 410 million tons (30.3%) were treated. Only 160 million tons (11.8% of total emissions) from polluting industry companies met emission standards, and a mere 67 million tons of treated effluent (16.2%) satisfied emission standards.

Air Pollution   Surveyed companies had an annual total of 6.2 billion cubic meters of air pollutants, of which 860 million cubic meters (13.7%) were treated in some manner. Emissions from production processes accounted for 3.9 billion cubic meters of the whole, and of those 1.2 billion (32.0% of production process emissions) were treated.

Solid Wastes   Polluting industry companies generated 49,260,000 tons of solid wastes, of which 15,990,000 tons (32.5%) were somehow treated. The recycled amount was 2,2070,000 tons, for a recycling rate of 44.8%.

Compliance with Environmental Regulations   EIAs were carried out by 22.7% of all surveyed companies, and 19.0% of polluting industry companies. Only 14.4% of all companies followed the "three simultaneous steps"[2] for taking the environment into consideration in new projects, a percentage that dropped to 11.6% for polluting industry companies. Payments of the pollutant emission surcharge amounted to 169,250,000 yuan for all surveyed companies, of which 111,330,000 yuan came from polluting industry companies.

Accidents, Shutdowns, Compensation, and Fines   There were 2,550 polluting accidents by all surveyed enterprises, of which 1,991 were by polluting industry companies. Companies shut down for pollution numbered 5,190, which is 5.74% of all shutdowns in China. Surveyed companies as a whole paid 21,030,000 yuan in compensation and fines, while the amounts for compensation and fines by polluting industry companies were, respectively, 14,430,000 and 1,740,000 yuan.

• Structural Problems

Based on these findings the survey report points out the following structural problems unique to TVEs.
(1) Priority to Economic Growth

Considered most important for TVEs is to increase employment and cash income. Thus, given the shortages of funds, raw materials, and energy, the environment gets short shrift as the companies pursue short-term increases in production and sales.
(2) Limitations due to Location and Size

Generally TVEs are small and at low technical levels, while pollution sources are geographically diverse, and the distribution of pollutants broad and rapid. In areas with low environmental capacity, such as those with small water endowments, pollution damage is locally very severe. Additionally, heavily polluting companies that find it difficult to operate in the cities have moved to rural areas.
(3) Inadequacies in the Environmental Management System

---

[2]   A rule that when designing, building, and operating a factory, one must at the same time design, build, and operate pollution control equipment.

Central government policy assigns low priority to rural environmental conservation, and levels of awareness and education regarding the environment are not high in rural areas. Further, rural areas lack environmental laws, pollution monitoring systems, and other systems for managing and conserving the environment.

• Case Study: Shanxi Province, Qianshan Coal Mine

In 1995 one of the authors worked with researchers from China's State Environmental Protection Agency on a study of the water resource management system in the Qianshan mining district in Shanxi Province.

This district lies near Taiyuan City in Shanxi Province, which is a major coal-producing area and water source for China. The breakdown of this area's 165 coal mines is four state-owned mines, three city mines, and 158 that are village-owned mines which fall into the TVE classification. Annual coal production by state-owned mines is 14,368,000 tons (72.8%) of the total, that of city-owned mines is 1,275,000 tons (8.9%), and that of village-owned mines is 2,635,000 tons (18.3%). Investigating mine shaft effluent from Qianshan coal mines showed that the average effluent index per ton of coal from mines classified as village-owned mines is far larger than that of state-owned mines (Table 1). Likely this is because state-owned mines have a certain level of production technology, while equipment at TVEs is neither well provided nor cared for.

On the basis of this study we can make the following observations on the management of coal mines and water resources.

(1) Research Data Not Released

Research in many areas including coal, geology, chemical industries, water utilization, urban construction, and environmental protection has for many years accumulated large amounts of data on the levels, amounts, quality, and hydrological conditions of water sources in the Qianshan mining district, but most of these data have been kept from the public in administrative offices. This is because departments consider these data to be administrative secrets, and do not share them with other departments. There is even a perception that releasing data would result in an encroachment on a department's turf.

(2) Pollution and Waste of Water

Hardly any effluent is treated, especially for recycling, so that wasteful water use is a constant occurrence despite water shortages. Qianshan coal mines have simple effluent treatment facilities in only three places, and only about 10% of the effluent is treated. Owing to insufficient funds, how-

ever, it is impossible to build effluent treatment facilities for the coal mining industry alone. It is also difficult to use money collected on the emission surcharge, which is in the hands of regional environmental protection agencies, because these funds are used only to install equipment in the factories that paid surcharges. Emission surcharge funds cannot be used for conserving the environment of an entire region.

(3) Complicated Management System

As the administrative divisions for resource management handle both water quality management and water development/use, water quality management suffers. And because 10-odd management organizations are involved in any one water source, the result is more chaotic than inefficient. In fact, many companies believed that as long as one has satisfied the conditions for formalistic EIAs, one would never be held accountable. Also, mines that lack funds tended to choose paying the emissions surcharge over investing in new equipment, and many individual mine operators deftly manipulate the chaotic management system to avoid paying even the surcharge.

(4) Lack of the Big Picture in Development

The chaotic management system means there is no consideration for the natural environment, such as water resources, when siting factories and coal mines. In recent years there are also problems that cannot be addressed under the rubric of headwater area alone, such as the disputes between headwater and demand regions over the conservation and development of water resources, and the growing sense of inequality in development cost burdens owing to regional differences in developmental stages. Yet in many instances there are no efforts for the planned conservation and use of resources from a perspective that sees the big picture.

Despite these problems of TVEs, not all of them are polluting, and many of them are more active than state-run companies in conserving the environment. Thus it is not the form of an enterprise, but rather its business management policy and profitability that determine how much it will invest in the environment. At any rate, China is in the process of institutional change, and with the exposure to foreign and domestic competition, there is a sharp increase in the number of companies with financial difficulties. Further, although bankruptcies used to be extremely rare, they are now a daily occurrence. The burden on the environment, meanwhile, will surely increase as fast as the number of TVEs.

2-3 Government Initiatives

China's environmental problems are changing with the times, and that change is always greatly affected by top-down politics and policy. Just as with economic policy, gov-

Table 1 Coal Production and Effluent Amounts of Qianshan Coal Mines

	Annual coal production (million of tons)	Annual effluent (million cubic m)	Effluent per ton of coal (m³/ton)
State-owned coal mines	10.46	8.65	0.83
Village-owned coal mines	2.64	5.97	2.21

Note: Totals for six state-owned and two village-run coal mines.

ernment policy on the environment develops through a trial-and-error process that is summarized below.

• Development of Environmental Policy

The origin of China's environmental policy is "integrated use," which began in the 1950s and consisted of afforestation campaigns and of resource conservation and recycling policy meant to alleviate resource shortages. In the 1960s and 1970s China maintained that there were no environmental problems under socialism, and while keeping the country half closed off from the outside world, the government further developed its policies on recycling and conservation, and widely publicized this as a spontaneous initiative. Building on such policy and the means to implement it, China has, since participation in the 1972 UN Conference on the Human Environment, developed an environmental policy and administrative apparatus modeled partly on that of the industrialized countries. These efforts involved trial and error, but were also subjected to interruptions and backtracking due to the political climate. At the same time, it was often the case that slogans and ideology took center stage, creating gaps with reality and lacking efficacy. But since about 1989, when the third National Environmental Protection Congress was held, policies and measures have quickly become more concrete. Factors behind this include the greater availability of information throughout society owing to reforms and opening of the country; worldwide concern for global environmental problems since the latter half of the 1980s; and trial-and-error efforts by the agencies responsible for environmental protection.

In autumn 1996 the government forced suspension or shutdown of several tens of thousands of projects and factories because of their pollution. Additionally, many environment-related laws were not only passed and enforced, but are now in the revision stage. In 1989 the Environmental Protection Law, which had been passed as an experimental law a decade earlier, was substantially revised and enforced as an official law. This was followed by the Law for Preventing Environmental Pollution by Solid Waste, the Noise Prevention Law, and others, plus revisions of laws including the Air Pollution Prevention Law (originally passed in 1987), and the Water Pollution Prevention Law (1984), making for the second round of environmental legislation. Revision of the Air Pollution Prevention Law changed the text from "preventing particulate pollution" to "preventing pollution caused by coal combustion." Additions to the law concern types of coal used and combustion methods, requirements for thermal power plants, and provisions to mitigate SO_2 and acid rain. Revisions to the Water Pollution Control Law included 16 new articles, and some of the new provisions are management and pollution prevention plans for rivers under the central or regional governments, protection of headwater areas, and penalties for violators, thus making it possible to widely implement the Huai River method. These changes are commendable as they build on the actual damage caused by pollution, and the policy successes in dealing with them. Nevertheless, an overview of the legal system reveals contradictions and inadequacies, and shows that owing to China's tradition of rule by man, rule by law has yet to win trust and permeate society. Under these circumstances it is sometimes the case that, in the course of solving problems, people use extralegal means to help themselves, or victims find they are helpless.

• Green Projects

The July 1996 the fourth National Environmental Protection Congress announced two major measures: the Total Pollutant Load Regulation Plan, and Inter-Century Green Projects. This congress, more than previous ones, emphasized the relationship between environmental policy and the five-year plans. The Ninth Five-Year Plan and Blueprint for Long-Term Objectives to 2010, which had been established in March 1996, mentioned total pollutant load control but not green projects. But by autumn of that year there was a substantially lengthened green project list of regions and items, and that November specific details were announced.

Green projects announced to date involve the simultaneous implementation of many priority environmental measures. Three phases cover the 15 years from 1996 to 2010. Phase 1 (to 2000) projects cover 1,591 items across a broad spectrum from local pollution problems to global environmental problems and administrative capacity building (Table 2). Over half the measures deal with water pollution, and most of them are concentrated in the East and South Central regions.

The Total Pollutant Load Regulation Plan tries to better mitigate pollution by adding total load control to the regulations that enforce environmental standards, which are based on pollutant concentrations to date. Under the Ninth Five-Year Plan, total load controls are imposed on 12 priority items: smoke, mining dust, SO_2, COD, petroleum products, cyanides, arsenic, mercury, lead, cadmium, hexavalent chrome, and industrial wastes.

• Environmental Investment

Under China's policy there are five mechanisms to obtain funds for environmental investments (here, mainly alleviating and preventing industrial and urban pollution). Percentages following the items are their approximate contributions.

(1) Three simultaneous steps investment (36%) (under the three simultaneous steps system)

(2) Renovation and remodeling investment (9%)

(3) Urban environment investment (39%)

(4) Pollutant emission surcharge investment (5%)

(5) Recycling profit investment (1%)

A comparison of actual investment with funds obtained shows that for China as a whole the compliance ratio for environmental investments is about two-thirds, with the remaining one-third going to environmental management personnel expenses and misappropriations by regional governments. If misappropriation was nonexistent and all funds went to environmental conservation, then according to a study by Li Zhitong, associate professor at Nagaoka Uni-

Table 2 Phase 1 Green Project Classification by Region

	North	Northeast	East	South Central	Southwest	Northwest	Other	Total
Water pollution control	73	80	332	214	61	35	6	801
Air pollution control	72	49	59	78	37	33		328
Solid waste control	41	24	49	46	20	21	71	272
Ecosystem protection	21	14	16	11	15	32	9	118
Actions on global environmental problems	7	2	4	6	4	7	39	69
National environmental capacity building							3	3
Total	214	169	460	355	137	128	128	1591

Note: Regions used in this table are based on Chinese administrative divisions, and differ from those used in this chapter. Regions here are defined as follows.
North: Beijing, Tianjin, Hebei, Shanxi, Inner Mongolia
Northeast: Liaoning, Jilin, Heilongjiang
East: Shanghai, Jiangsu, Zhejiang, Anhui, Fujian, Jiangxi, Shandong
South Central: Henan, Hubei, Hunan, Guangdong, Guangxi, Hainan
Southwest: Sichuan, Chongqing, Guizhou, Yunnan, Tibet
Northwest: Shaanxi, Gansu, Qinghai, Ningxia, Xinjiang

versity of Technology, it would be possible to meet the Chinese government's environmental investment goal of 1% GNP (from 1991 to 1994 it was about 0.8%).

About 60% of China's environmental investment goes to urban sanitation, recycling, tap water, refuse collection, and the like under management at the provincial or city level, while about 20% goes to factories and facilities under the management of industry-related management organizations (such as the Chemical Industries Ministry) or the military. Direct investment by foreign companies, investment by the World Bank and other international organizations, and that under bilateral aid accounts for the remainder.

This "micro" look at China's environmental problems in relation to region, TVEs, specific policies, and fund procurement mechanisms shows that the causes and remedial measures for environmental problems differ considerably according to region and company. Yet even this vast and varied country is but one of the many members of global society. The following section is a "macro" look at how China and global society as a whole affect each other.

3. The Global Environment and China

3-1 Impacts on Global Society

• Global Warming: Cause and Victim

China could very well be a major victim of global warming, but it might also be a major cause. A sea level rise would mean heavy damage to coastal cities such as Shanghai and Tianjin. The Intergovernmental Panel on Climate Change predicts that a one-meter sea level rise would affect 70 million Chinese, a figure equal to that for India, and the highest in the world. It is certain that weather extremes induced by climate change will bring about drought and flooding throughout China. Agricultural production will be greatly affected, according to predictions by Japan's National Institute for Environmental Studies (NIES) and Nagoya University, which say that a 2.5°C rise in air temperature will translate into a 40% decrease in Chinese corn production.

But China is also a major cause of global warming because of its huge greenhouse gas (GHG) emissions. Presently China's per capita CO_2 emissions are only about one-fourth those of Japan and one-tenth those of the U.S., but on a by-country basis they are second after the U.S. International Energy Agency predictions say that the increase in CO_2 emissions by China alone from 1990 to 2010 will equal the increase by all OECD countries combined.

Internationally there is much discussion on how to deal with global warming, but in China most discussion centers on the need to conserve energy. In the international arena China notes the responsibility of industrialized countries, but, perhaps because of concerns that it will be forced to cut GHG emissions, it is very sensitive to the contention that China's future emissions will be the deciding factor in global warming.

• Energy Supply and Demand

Calculations by Japan's Keio University and the Ministry of International Trade and Industry (MITI) show that if China consumes energy at the rate of semi-developed countries and at its present efficiency, China alone will account for 80% of world energy demand. While China's energy efficiency has improved in recent years, the increase in oil consumption brought about by higher automobile demand alone means the situation is serious. The Institute of Energy Economics in Japan estimates that China's present motor vehicle fleet will increase from the current 10 million vehicles to 40 million in 2010, which would make China's 2010 road transport oil consumption about 150 million tons, or 1.4 times that of Japan.

China promotes nuclear power as one chief way of dealing with growing energy demand. Currently there are three plants in operation, at Qinshan in Zhejiang Province and Daya Bay in Guangdong Province, and plans call for building eight plants in four sites by 2000. And while a prime concern about nuclear power is safety and the risk of accidents, China's technical level in this area — a decisive factor — is not believed to be as high as that of industrialized countries. But China has increasingly opened itself to the outside world: it has joined both the International Atomic Energy Agency and the Nuclear Non-Proliferation Treaty, and has entered into agreements with 10-odd countries for the peaceful use of nuclear power. It is also a signatory to the Convention on Nuclear Safety and the London Dumping Convention. But China has inadequate nuclear power-related legislation, no law for nuclear accident compensation, and is not a signatory of the Convention for Civil Liability for Nuclear Damage. China's risk management system also lacks a mechanism for considering citizen and media criticism.

Growing energy demand in Asia means many countries will further increase their dependence on Middle Eastern oil, which will further aggravate territorial issues (such as the Spratly and Senkaku Islands) involving security and resources in the South China Sea, the route by which Middle Eastern oil comes.

3-2 Transboundary Acid Rain

• The Challenge of Chinese Emissions

Although massive desulfurization investments by Japan in the 1970s substantially cut emissions of acid rain precursors, fiscal limitations in China make it highly unlikely that installation of flue gas desulfurization equipment and other hardware will quickly rise from the current several percent. With a per capita GDP only one-fiftieth that of Japan, to China it makes much more sense to alleviate its acute energy shortage by building another new power plant than to outfit an existing plant with costly desulfurization equipment. For the time being, China must depend on the cheap coal that is readily available domestically. Even if China could afford the switch to low-sulfur oil as Japan has done, that would mean increasing oil imports, which would raise the international price and have a serious impact on international society.

• Sulfur Oxides from China

Before getting into a quantitative discussion, it is useful to see how China-related acid rain became an issue in Japan. It was in 1982 that Japan's Environment Agency set up a panel to assess acid rain measures, and in 1985 it performed a survey with the Forestry Agency on damage to cedar trees and the connection to acid fallout. In 1993 NIES found that when westerly winds blow over the Sea of Japan off the coast of Shimane Prefecture there are higher concentrations of sulfur and nitrogen oxides, the causes of acid rain, which demonstrated that acid rain precursors come to Japan from the Asian mainland. A report released by the Environment Agency in 1994 was the first official observation that SOx of continental origin were affecting Japan.

Quantitatively gauging the deposition of acidic substances is not easy and yields large errors. However, a number of Japanese researchers estimate that several percent of China's sulfur emissions are transported to Japan, and that sulfur from China accounts for 10 to 50% of sulfur deposition in Japan (Fig. 1).

In Europe there is an overall qualitative and quantitative consensus on the sources of acid rain. By contrast, while Japanese and Chinese researchers are fairly certain that to some extent sulfur from China is transported to Japan and deposited, there is no adequate consensus between the two countries on the amount, and the extent to which it causes acid rain in Japan. Another difference between East Asia and Europe is the regional difference in critical load. While damage is severe in some parts of Europe, research shows that Japanese soils have a high acid neutralization capacity. And because in Japan volcanic activity accounts for perhaps 20% of acid precipitation-causing sulfur deposition, this imposes a limit on human efforts toward reductions.

• Can Japan Help China?

In view of the foregoing, Japan would find difficulty in claiming that it is at this time subjected to severe transboundary pollution. What is clear, however, is that in the near and mid term there will be no change in China's dependence on coal and in the low priority accorded to desulfurization. Even if only a small part of China's emissions reach Japan, provided that total Chinese emissions and their rate of increase are large, Japan's situation could become serious in the future.

Another consideration is the gravity of damage in China. For example, a 1992 study in Chongqing City quantified the serious harm to human health, crops, and buildings. The morbidity rates of chronic rhinitis and sore throat were several times higher in heavily polluted areas than areas not so polluted (Table 3). Central Chongqing's lung cancer mortality rate jumped more than twofold from 1973 to 1989 (Table 4). Further, the city's Institute of Environmental Sciences estimates that in health damage alone this pollution is costing 231,670,000 yuan per year (Table 5).

It is to be hoped that Japan's world-class desulfurization technology and experience with administrative-level desulfurization measures can be used in joint Chinese-Japanese development of technologies with improved cost performance, but this will necessitate more research and development for environmental technologies by not only business, but also public research institutes.

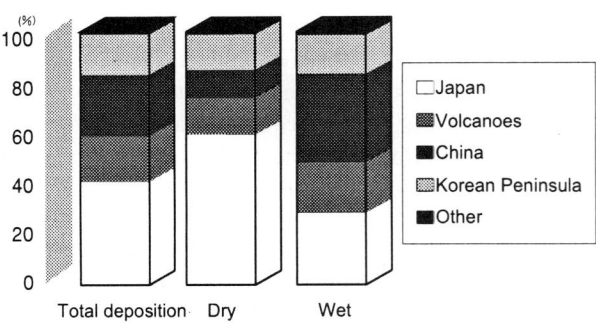

Fig. 1 Sources of Sulfur Deposition in Japan and Their Proportions

Source: Central Research Institute of Electric Power Industry.

Table 3 Cases of Chronic Rhinitis and Sore Throat in Three Areas of Urban Chongqing

	Heavily polluted			Fairly heavily polluted			Not seriously polluted		
	No. of subjects	No. of patients	Morbidity (%)	No. of subjects	No. of patients	Morbidity (%)	No. of subjects	No. of patients	Morbidity (%)
7-year-old boys	130	18	13.8	101	3	3.0	109	2	1.8
12-year-old boys	36	9	25.0	66	7	10.6	59	3	5.1

Note: Study performed 1989-1990.
Source: Chongqing Institute of Environmental Sciences.

Table 4 Lung Cancer Mortality in Central Chongqing (1973-1989, 1/10,000)

Year	1973	1978	1983	1984	1986	1987	1988	1989
Mortality	21.8	45.8	52.4	42.2	51.6	49.1	50.4	54.7

Source: Chongqing Institute of Environmental Sciences.

Table 5 Economic Loss to Acid Damage in Chongqing

	Ecosystem	Human health	Artificial structures
Difference in damage between acidified areas and others	25% drop in vegetable and grain production; 300 sq. km of forests die.	Doubling of lung cancer deaths; double rate of respiratory ailment sufferers; four times per capita medical costs.	Metal corrodes eight times faster; paint lasts one year (under half the time of ordinary areas).
Monetary loss due to acid damage (annual)	Grains 4,790,000 yuan; vegetables 28,730,000 yuan; forests 35,100,000 yuan.	Medical costs per patient about 90 yuan; total 231,670,000 yuan including labor loss.	Metal materials 35,100,000 yuan; total 125,810,000 yuan including paint expense loss.

Source: Chongqing Institute of Environmental Sciences.

3-3 Building an Environmental Security System

• Current China-Japan Environmental Cooperation

Let us divide cooperation into the following five categories and discuss each briefly. (1) Official development assistance (ODA), (2) Green Aid Plan, (3) local governments, (4) the private sector, and (5) NGOs.

ODA In both policy and in substance, ODA is changing from its former emphasis on economic infrastructure to one on conserving the environment. Fifteen of the 40 items under the fourth yen loan (1996-2000) to China were environmental, and some were implemented ahead of schedule. Most of the environmental items in China's original proposal remained in the final proposal. Probable reasons for attaching importance to the environment in yen loans are because China now gives the environment higher priority, and because of financing policy, i.e., environmental projects with low direct economic benefit get low-interest yen loans, and infrastructure construction with large direct benefits gets World Bank and private-sector financing.

Green Aid Plan A MITI aid program in place since 1992, its purpose is to transfer environmental technologies to developing countries in Asia and foster environmental industries (its FY1998 budget was 13.9 billion yen). Much of its aid consists in contributing model environmental equipment plants for air pollution or water treatment, and over half the projects are for China. It is more flexible than ODA, and it is easier for Japan to take the initiative. While the Green Aid Plan provides technical cooperation as ODA does, a significant challenge is how to work around fiscal and institutional limitations so that technology transfer from Japan leads to the dissemination of technologies within China.

Local governments Former heavily polluted Japanese cities such Kitakyushu (sister city of Dalian), Yokkaichi, and Kawasaki (sister city of Shenyang) conduct interpersonal exchanges and transfer expertise in dealing with pollution. Such cooperation emphasizes knowledge and know-how over hardware, and is capable of sustained efforts using local industries. Japan's government has begun providing financial support to such international environmental cooperation by local public entities.

Private sector Basically companies are most concerned with profitable activities, and it goes without saying that transferring environmental technologies in conjunction with expanding private-sector investment is most efficient and likely to continue. In reality, however, there are still few instances of direct investment in which Japanese companies manufacture environmental equipment in China, reasons being that demand in China is still small, and that it presents considerable risks for Japanese companies. Expanded public support is therefore quickly needed.

NGOs Cooperation consists mainly in interpersonal exchanges by way of activities in China such as nature conservation and public education. In 1993 the Japan Environment Corporation initiated a system that provides NGOs with financial help, and nature protection work such as afforestation accounts for a large proportion of China-related NGO activities benefiting from this system.

• Pollutant Reduction Agreement and Incentives Needed

Many parties are involved in Japan-China environmental cooperation, which grows year by year. But current circumstances make it hard to take the acid precipitation threat and global warming seriously: In Japan there is no sense of imminence about these problems, and energy prices have fallen back to their pre-oil shock level, while at least in the developed countries and parts of Asia there is material plenty. Nevertheless, as the grave ecological impacts of acid precipitation and global warming become evident, controls on atmospheric emissions of sulfur and CO_2 will be indispensable to provide stronger incentives for cooperation and business efforts. The first step toward creating such systems would be an emission reduction agreement on regional pollutants, with an example being the International Treaty on Long Range Transboundary Air Pollution.

But Asian countries have yet to formulate concrete policy due to the lack of basic information on sulfur deposition amounts, emissions, long-range transport, and ecosystem impacts. An Acid Deposition Monitoring Network in East Asia is now being built as a Japanese initiative, but it was only as recently as 1996 that a decision was made on guidelines for methods to measure sulfur deposition, and usable data will not be available until 2000 or later. Asia also has no organizations such as the European Union or the Conference on Security and Cooperation in Europe to consolidate talks on security matters. In terms of the infrastructure for scientific research and political action, Asia is unfortunately 10 or 20 years behind Europe.

• Building an Environmental Security System

Nevertheless, there have been changes over the past decade, with worldwide discussion on flexibility mechanisms such as the clean development mechanism (CDM), joint implementation (JI), and emissions trading. Under an ideal scenario, schemes employing the market mechanism would lead to the organization and institutionalization of international cooperation aimed at building an environmental security system. For example, if the CDM and JI are to work as a system, they would need the cooperation of the academic community, such as in performing cost-benefit analyses on the basis of a uniform format. Also needed would be an international monitoring organization, and greater interchange among businesses. Working together more and expanding channels of communication would strengthen the sense of regional unity and create a feeling of trust. And if the CDM and JI succeed in bringing about technology transfer, both giver and receiver will of course benefit.

To China there is great significance in the transfer of advanced technology, which would not have happened without the CDM. And if joint research on costs and benefits demonstrates the compound benefits conferred, such as the energy savings realized both in China and other countries by reducing air pollution emissions, China too will come to have a high estimation of such programs and environmental technologies themselves. Asia needs to build an environmental security system, with long-term policy objectives, that advances regional cooperation in a more organized manner and creates relationships of trust as it considers also the stockpiling of oil and food, nuclear power risk management, and other problems.

4. Summation

China is a land of the old and new — of expressways and oxcarts, of computers and abacuses. In its long history it has been through empire, semi-colonization, and socialism. China's environmental problems are as vast and varied as its natural landscape, which makes a complete understanding nearly impossible. This is especially so to people in the developed countries, who have little direct sense of energy shortages and other problems in the developing world. But China's problems are real, and they are going to affect everyone. Such being the case, we need a policy of engagement toward China. Merely calling attention to China's responsibility for consumption or pollution would be taking the easy way out; needed is perseverance in building a framework for cooperation.

(ASUKA Jusen, JIN Song, AIKAWA Yasushi)

Essay 1   Environmental Studies Flourish while the Environment Languishes

China already has the world's largest number of environmental scholars, who busy themselves with learning about the environmental research, laws, and policies of the developed countries. No other developing country government puts as much effort into the environment as China. Its constitution has a provision for environmental conservation, and population control is one of its basic national policies. State-owned companies have achieved marked improvement in managing heavy metals and solid wastes, and the government carried out with drastic measures by closing down 60,000 highly polluting small companies by administrative order.

Yet, as the irony in the title's parody of the famous classical poem by Du Fu suggests, pollution keeps worsening, and China has the worst pollution of anywhere in the world, thanks to the rapid percolation of material civilization into this country. But China has yet to come up with new institutions and values to control this new civilization. The world's largest proportion of heavy and chemical industries, along with a very high proportion for the mining industry, produce vast amounts of mineral poisons, slag, and other wastes.

Developed countries curbed pollution by spending over 1% of GDP on the environment, but China will probably need twice that figure, requiring it to take a hard look at its spending priorities.

(KOJIMA Reietsu)

Essay 2   Cutting Waste and Pollution

It is said that China uses 30% to 80% more energy to produce the same product as in developed countries. That means energy is being wasted, but materials are of course being wasted, too. And that wastefulness translates into increased generation of wastes. Essential in cutting pollution is stopping it at the source, which is the workplace, and that necessitates reducing the waste of energy and materials. But herein lies the business management flaw of Chinese companies. Lax production management raises product costs and puts many state-owned companies in the red. In China company deficits, energy and material waste, and pollution have the same causes.

In the past Japanese companies also polluted badly, but the cause was insufficient investment in waste management facilities, not poor production management. This experience too could help if China learns from it.

(HOSHINO Yoshiro)

Essay 3   Another Look at the Food Crisis Controversy

Who Will Feed China?, a 1994 book by Lester Brown at the Worldwatch Institute, stirred up more controversy than at any time since the publication of *Limits to Growth* by the Club of Rome in 1970. Brown's point was that despite China's growing appetite, its food production will decrease substantially because of pollution and changes in its industrial and social structures. Many tried to refute that argument by claiming that technological innovation and government policy will increase both yields and land in cultivation. Both optimists and pessimists are using projections for several decades into the future, whose accuracy is of course open to question. Still, there are three items in the arguments of the Chinese government and agriculture specialists — who argue optimistically — which call for special attention.

First, the optimists are not claiming that nothing need be done about the present situation. Second, although optimistic scenarios assume technological advances and help from developed countries such as funding and technology transfer, such transfers are not making as much progress as hoped. Third, there are always negative concomitants when using more advanced agricultural technologies. For instance, in 1994 the use of pesticides saved 42,770,000 tons of food, but it is estimated that about 100,000 people suffer from pesticide poisoning each year, and in 1994 nearly 10,000 of them died.

Another consideration is that to many Chinese, U.S. statements on China's food and population present a double standard because while the U.S. criticizes China on human rights grounds for its population reduction measures, China is considered a threat because its population increase results in more food consumption.

One hopes that Brown's book will not be taken in this way, but rather as a call for concerted international action to help developing countries.

(ASUKA Jusen)

Chapter 7
Taiwan

1. Introduction

In the 1960s and 1970s Taiwan had average GDP growth of nearly 10% in real terms. Spectacular economic growth continued even after that, raising the per capita GNP of about $100 in 1950 to over $10,000 in 1992. Such rapid industrialization on this small land area necessarily weighs heavily on the environment, and Taiwan's environmental burden per unit land area attained the world's highest level. Countries that pursue accelerated industrialization must from the initial stage of economic growth make environmental policy efforts commensurate with the rate of growth, but Taiwan's efforts have been inadequate. In fact, Taiwan's unique difficulties emerged upon giving environmental conservation a place in public policy. Generally development that accords priority to catching up with other countries promotes the highly polluting heavy and chemical industries over others. For this reason, policy tends to favor industrial infrastructure over livelihood-related community development, to overemphasize education in engineering and the sciences for developing human resources, and to encourage concentration in large cities, a course of action that readily causes environmental damage. For the nearly four decades from May 19, 1949 to July 16, 1987 Taiwan's state of martial law impeded the functioning of democratic channels, which play a vital role in getting the environment onto the policy agenda. Under martial law the people developed feelings of distrust and discontent toward government and business for causing pollution and environmental problems, and when martial law was lifted, those feelings erupted into the protests and petition drives known as "self-help," which shall be discussed below. Advances in democracy gave environmental concerns a place on the political agenda, and Taiwan is building its system of environmental law and policy. At the same time, the global economy presents Taiwan with rapid economic and industrial structure changes, and presses it to move in a new direction. This chapter will give a general account of Taiwan's environmental problems, and then focus mainly on the self-help movement while discussing and analyzing the process by which environmental problems gained social recognition and were formally integrated into policy.

2. Environmental Problems in Taiwan

2-1 Air Pollution

Taiwan's government determines the state of pollution with its Pollutant Standards Index, or PSI, which is calculated by measuring the concentrations of airborne particulates (PM_{10}), SO_2, NO_2, CO, O_3, and other pollutants on a certain day, converting their measured values to sub-indexes, and using the maximum values for that day's PSI. Their evaluations are: (1) 0<PSI<50, good, (2) 51<PSI<100, normal, (3) 101<PSI<199, bad, (4) 200<PSI<299, very bad, and (5) 299>PSI, hazardous.

As of 1995 Taiwan had a total of 232 air pollution monitoring stations, which comprise 66 automated stations directly run by the central government's Environmental Protection Administration, 14 automated stations run by the Departments of Environmental Protection of counties, cities, and other local bodies, and 152 manually operated stations. The PSI for the 12-year period from 1984 to 1995 was, between the years 1984 to 1991 for which data are available, "bad" or "very bad" on about 15% of the days, but the figure for 1995 was 6.10%, indicating a trend toward gradual improvement. A comparison of actual air pollutant emissions between 1988 and 1995 (Table 1) shows that of the main pollutants, the emissions of total suspended particulates, SOx, CO, and total hydrocarbons (THC) decreased 48.6%, 48.8%, 21.2%, and 28.2%, respectively, during that interval. For that reason the rate at which the PSI topped 100 declined.

Pollutant emissions decreased because of emission controls. Beginning in September 1989 Taiwan conducted the "Combined Land-Air Investigation of Stationary Air Pollution Sources," in which airborne police and inspectors from the Environmental Protection Administration track pollution sources from the air in helicopters, while inspectors from the Adminstration who are waiting on the ground visit the pollution sources to initiate control actions. By June 1995 these inspections had discovered a total of 8,778 instances of illegal smoke emissions at factories or open-air burning, of which 648 (7.4%) were formally charged. Beginning on April 12, 1990 the authorities conducted the "Plan for Inspecting Stationary-Source Smokestack Emissions of Air Pollutants," which as of June 30, 1995 had inspected a total of 3,164 factory smokestacks, of which 581 (18.4%) failed. Smokestacks that failed were charged with violations, fined, or subjected to other administrative actions, and ordered to effect improvements by a certain deadline. These actions and regulations have resulted in a trend toward gradual mitigation of air pollution, but this does not hold for NOx, many of whose emissions are from motor vehicles. Under the Air Pollution Fee Collection Law, which was promulgated on March 23, 1995, Taiwan began collecting a kind of air pollution tax on July 1 of that year. Although there is controversy over the efficacy of this "tax" and the way its revenues are used, it is clear that Taiwan's government is making positive use of economic means as a way to incorporate consideration for the environment into the market economy framework.

In 1995 the major emission sources of air pollutants except for volatile hydrocarbons could be classified in three groups: (1) Transportation, including private automobiles, gasoline- and diesel-powered vehicles, and two-wheeled vehicles, (2) industrial production, and (3) combustion, comprising industry, power generation, and open-air burning. Overall, transportation accounts for about half, excluding THC and SOx, with slight differences among pollutants. Air pollution is a serious and growing problem because of increased demand for transportation in conjunction with industrialization and urbanization. Emissions of SOx and NOx

Table 1 Comparison of Air Pollutant Emissions, 1988 and 1995

	a. 1988 emissions (t/yr)	b. 1995 emissions (t/yr)	c. Change = (b-a)/a
Total suspended particulates	1,266,899	650,688	-48.6%
Airborne particulates (PM$_{10}$)	-	399,437	-
Sulfur oxides (SO$_x$)	869,907	445,460	-48.8%
Nitrogen oxides (NO$_x$)	638,591	646,103	1.2%
Carbon monoxide (CO)	1,891,213	1,489,543	-21.2%
Total hydrocarbons (THC)	874,371	628,018	-28.2%
Non-methane hydrocarbons (NMHC)	-	607,649	-
Lead	-	319	-

Source: Prepared from the 1990 and 1996 editions of *Yearbook of Environmental Information, Taiwan Area, the Republic of China.*

from thermal power plants account for 50.8% and 21.5% of their totals, respectively, while their figures are 24.6% and 8.8% from industrial combustion.

2-2 Water Pollution

About 70% of Taiwan Island is mountainous, and less than 30% of it is level land below 100 meters elevation. Slopes are steep and rivers are swift. Rivers on the western plain, where the population, cities, and industrial zones are concentrated, run mostly east to west, making for short rivers with narrow watersheds. The longest river in this region is the 186-km Choshuichi. These geographical conditions mean that despite Taiwan's plentiful annual rainfall, only about 21% of it can actually be used, and owing to rapid population growth and economic development, use of especially industrial and household water tends to increase every year. Because almost all of this water comes from rivers, dam reservoirs, lakes, and groundwater, the environmental burden is very heavy.

Taiwan Island has 21 class 1 rivers, 29 class 2 rivers, and 79 ordinary rivers. Environmental offices with jurisdiction over these rivers check their quality periodically. A river's state of pollution is determined by its dissolved oxygen, biological oxygen demand, suspended solids, and ammonia nitrogen (NH$_3$-N), and rivers are classified into one of four pollution levels: (1) unpolluted, (2) minor pollution, (3) moderate pollution, and (4) serious pollution.

Water quality test results for Taiwan's class 1 and 2 rivers in 1995 (Table 2) show that only 64.2% of rivers are unpolluted, and that 35.8% of Taiwan's primary rivers are polluted to varying degrees. Additionally, river pollution tends to worsen as indicated by the unpolluted proportion of

73.1% in 1986 declining to 64.2% in 1995. River pollution in turn induces more groundwater pumping.

Eutrophication indexes (Carlson TSI) for Taiwan's main lakes and dam reservoirs from 1993 to 1996 show that of 24 bodies of water, 13 (54.2%) were eutrophic in 1993, 12 (50%) in 1994, four (16.7%) in 1995, and six (25%) in 1996. The worst eutrophication was in the southern dam reservoirs of Akungtien, Cheng-ching-hu, and Fengshan. In fact, water at Akungtien and Fengshan was switched to industrial use because quality was below that for drinking water.

Groundwater is likewise contaminated. Non-compliance ratios for groundwater quality test categories over the 13 years from 1983 to 1995 indicate a very high average of about 30% non-compliance for the biological contaminants of coliform bacteria and general bacteria. Tests have also continually detected lead, arsenic, and other toxic substances, as well as nitrate nitrogen.

Results of test items for pollutants that affect drinking water are also less than desirable, as groundwater contamination is widespread and has serious consequences. And unlike air pollution, water pollution shows no signs of improvement at all. Water pollutant emission sources can be roughly divided into the three categories of domestic wastewater, industrial effluent, and livestock effluent. Wastewater from these sources flows directly, or through underground channels, into rivers throughout Taiwan with hardly any treatment, thereby polluting the mid and lower reaches of rivers. Although stiffened environmental regulations promise some degree of improvement in pollution from industrial effluent, livestock effluent, and other sources, there is still little hope for mitigation of domestic wastewater because many areas still have no sewerage, and construction will take time. Additionally, as of December 1996 Taiwan had about 11

Table 2 Pollution of Primary Rivers, 1995

	Class 1 rivers (21)		Class 2 rivers (29)		Total (50 rivers)	
	Km	%	Km	%	Km	%
A. Unpolluted	1,297.70	62.14	584.90	69.20	1,882.60	64.16
B. Minor pollution	262.89	12.59	75.00	8.80	337.89	11.52
C. Moderate pollution	205.06	9.82	114.50	13.60	319.56	10.89
D. Serious pollution	322.66	15.45	71.30	8.40	393.96	13.43
Totals	2,088.31	100.00	845.70	100.00	2,934.01	100.00

Source: Prepared from the 1996 edition of *Yearbook of Environmental Information, Taiwan Area, the Republic of China.*

million pigs and 98 million chickens. Assuming that one pig excretes four times as much as one human, Taiwan's pig population creates a serious environmental burden on the water supply. Many pigs and chickens raised in Taiwan are exported to Japan.[1]

Pesticide application per unit cropland in Taiwan is very high (Fig. 1). Owing to the warm climate, most farmland yields two crops a year, and some places in the south have three. Frequent pesticide applications affect cropland fertility, but residual chemicals also seep into the soil and wash out with rainwater, contaminating groundwater and the water stored in lakes and dam reservoirs. The sewerage connection rate in Taiwan is internationally very low (Fig. 2). Even though per capita national income is already over $10,000, only 3% of Taiwanese are provided with the sewerage services so essential to urban life, and almost all of those 3% live in Taipei. The mayor of Taipei, who was elected in December 1993, declared that he would raise Taipei's sewerage connection rate to 40% within his term.

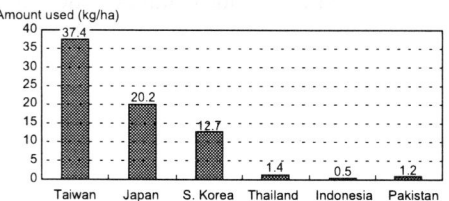

Fig. 1 Pesticide Application per Unit Area of Cropland

Source: Prepared from *Environment Taiwan — Contributing to Taiwan's Future*, 1996.

2-3 Soil Contamination

One factor that induced Taiwan to begin soil contamination surveys was the discovery in the mid-1970s of cadmium-contaminated rice in Taoyuan County, northern Taiwan, a discovery which confirmed that soil contamination had made its way into agricultural produce. A survey of Taiwan's topsoil (0-15 cm) for the three elements of arsenic,

---

[1] Other authors have demonstrated how other exports to Japan, such as shrimp, also impact heavily on the producer countries' environments.

cadmium, and copper revealed that, notwithstanding differences in extent among cities and counties, the contamination is widespread. Especially in Yunlin County in central Taiwan, and in Tainan and Kaohsiung counties in the south, large areas are contaminated by arsenic. Despite this, Taiwan has no law that controls cropland soil contamination by specifically designating certain chemical elements.

In Taiwan the organizations that manage farmland and irrigation water are called "Water Conservancy Associations," which number 13 in all. "Instances of serious pollution" registered with each association (Table 3) are caused by chemical plants, heavy metal contamination by plating plants, urban wastewater, aquaculture effluent, and other sources, and many of them have caused soil contamination either directly or indirectly. Especially the Lichangjung Chemical Plant in the Hsinchu area caused widespread damage through continued water and air pollution, which elicited displeasure and protests from citizens, and led even to four "self-help" incidents between 1982 and 1988 in which citizens surrounded the plant. The causes of such soil contamination can be generally summarized into the following five types.

First is contamination by chemical fertilizers and pesticides. During the five years from 1990 to 1994, each of Taiwan's farmers used a yearly average of about 37.4 kg of both. Excessive use of these substances directly jeopardizes human health, and also decreases soil fertility and pollutes water.

Second is the lack of action on heavy metals from industrial production processes. When untreated heavy metals are left on the ground they are absorbed by the soil or oxidized. Elemental metals and their oxides are gradually washed away by rain, spreading the pollution farther. Heavy metals have contaminated about 50,000 ha, which is larger than the combined area of Taipei and Kaohsiung cities.

Third is untreated industrial effluent, which contaminates soil when discharged directly onto the ground, and indirectly contaminates farmland when polluted river water is used to irrigate crops.

Fourth is landfill wastes that contain pollutants. When disposal site management is inadequate, pollutants in buried wastes are sometimes directly absorbed by the soil.

Fifth is acid rain, high rates of which are found in Taipei,

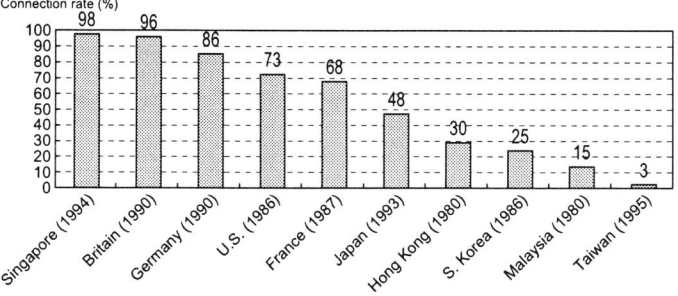

Fig. 2 Sewerage Connection Rates for Various Countries and Regions

Source: Prepared from *Environment Taiwan — Contributing to Taiwan's Future*, 1996.

Table 3 Examples of Serious Pollution of Cropland for Each Water Conservancy Association

Association	Recent example
Taipei, Keelung	Heavy metal contamination by Taikuo Metal and Chemical Plant
Taoyuan	Heavy metal contamination by Kaoyin and Chili chemical plants
Shihmen	Contamination of farmland by Chiahsin Farm, factories along the Laochieh River, and urban effluent
Hsinchu	Water and air pollution by the Lichangjung Chemical Plant
Miaoli	General pollution of Chunan and Toufen areas
Taichung	Pollution of Wu River source, effluent from plating plant, urban wastewater pollution of Pabao River and Hulutun
Changhua	Pollution by effluent from illegally operating plating plant
Yunlin	Water pollution by effluent from mid- and small-scale hog farms and Hungkang paper mill
Chianan	Irrigation suspended due to water pollution in Sanyehkung River
Kaohsiung	Water source pollution in Houching River irrigation zone and Erhjen River; pollution of Fengshan and Neiho rivers by urban effluent
Pingtong	Pollution by Pingtong Paper Mill effluent, pollution by wastewater from shrimp farms in vicinity of Chiatung and Linpien
Hualien	Pollution by effluent from marble industry
Ilan	Contamination of cropland by brine from shrimp culturing ponds

Source: Prepared from the 1996 edition of *Yearbook of Environmental Information, Taiwan Area, the Republic of China.*

Kueishan, Chungli, and other northern locations, as well as Hsiaokang in the south.

2-4 Noise

The Environmental Protection Administration's 1995 statistics for noise complaints categorized by source show that factories accounted for 33.62% of complaints, followed in descending order by entertainment businesses, construction sites, broadcasting facilities, and neighborhood noise at 28.78%, 16.35%, 11.22%, and 10.29%, respectively. The primary cause of noise problems is zoning. Siting residential developments near or within commercial and industrial zones results in frequent mixing of residences with businesses and industries, which is the basic cause of noise complaints. Taiwan has an especially large number of family-run factories that operate around the clock on three shifts, as well as entertainment businesses open late at night near residential areas, so that complaints about factories and entertainment businesses together account for 60% of the total. Complaints about traffic noise are unexpectedly few, the reason probably being that people often simply give up on lodging complaints because vehicles are always on the move and soon disappear.

To deal with the constantly increasing number of noise complaints, Taiwan implemented the Motor Vehicle Noise Control Law on January 1, 1991, requiring that manufacturers and importers of all motor vehicles submit for each model a motor vehicle inspection certificate that has been obtained by having a sample vehicle tested and approved by the government. A law to control aircraft noise was promulgated on August 31, 1994. Taiwan also installed 59 noise monitoring stations around six airports to measure aircraft noise, and 19 stations around the country to measure road noise. While these stations are concerned primarily with transportation noise, a law called Environmental Loudness Standards was finally promulgated on January 31, 1996 to control the

"five major noise sources" of factories, entertainment businesses, construction sites, broadcasting facilities, and neighborhood noise. It remains to be seen how effective these measures are.

2-5 Ecosystem Damage

At the end of the 19th century nearly 80% of Taiwan was forested. The alpine regions of the Ali Mountain Range alone had about 1.5 million trees in old growth forests having species characteristic of Taiwan. Indeed, Taiwan had a rich store of old growth forests, but these forests of 1.5 million trees were almost completely logged by the mid-1960s. Full-scale logging of Taiwan's forests began in 1895, when it was a Japanese colony. Of the approximately 6 million cubic meters of old-growth forest in the Ali Mountain Range, about 4 million were temperate and cool-temperate mountain conifers with high economic value, and 2.5 million of those were logged in the 50 years of colonial domination, meaning that in that half century Japan logged about 13% of Taiwan's land area. The remaining 1.5 million cubic meters were completely logged in only 15 years by Taiwan's Department of Forestry, which further accelerated the loss of forest resources.

In 1945 just after the war ended, forests covered about 65% of Taiwan, but that fell to 62% by 1971 and to 52% in 1994. Over the last 50 years Taiwan has lost 420,000 ha of forests, which is about 15 times the land area of Taipei. Activities by private organizations to protect forests began about 1986, with the biggest achievement being that in August 1991 they made the Department of Forestry stop its excessive logging.

But in the 1960s a new phase of development had begun to rapidly diminish forest resources. Because urbanization had quickly consumed all the flat land available for development in postwar Taiwan, new land was in short supply despite the increasing demand generated by the growing

114

population and industrialization. This situation pushed development toward the mountain forests. In the early 1960s Taiwan's future president Chiang Chingkuo led 100-odd former military men to a 1,500-meter mountain called Lishan in central Taiwan and started planting apple trees, which was a strong inducement leading to never-ending reckless development and opening of land in Taiwan's mid-elevation mountainous areas. As of 1996 about 700,000 ha had been developed with recreation facilities, as well as tea plantations, orchards, betel palms, and other cash crops, a total area about 1.7 times that of the forests lost in the postwar years. Because such development violates land use restrictions it should have been strictly curbed, but as the authorities have let developers do as they please for such a long time, it has become a fait accompli. A more serious problem is that this illegal development not only destroys forest resources, but also makes dam reservoirs eutrophic and pollutes rivers because developed land, being all located on the upper and mid reaches of rivers, and above dams and other water supplies, allows residual pesticides, soil, and the like to be washed downstream.

Taiwan island has a large number of plant species, with vascular plants alone accounting for about 4,000 species, and not a few of them are endemic to Taiwan. Each old growth flora community is home to different wild animals, resulting in a variety of ecosystems. The cool-temperate and warm-temperate mountain coniferous forests rapidly cut during the Japanese colonial era not only had many of Taiwan's endemic flora, but also at least 40 snake species, over 150 alpine bird species (13 endemic), and three frog species, plus the Snow Mountain grass lizard, Formosan salamander, Formosan sambar (*Cervus unicolor swinhoei*), Chinese river otter (*Lutra lutra*), Formosan serow (*Naemorhedus swinhoei*), Formosan black bear (*Ursus thibetanus formosanus*), Formosan macaque (*Macaca cyclopis*), Formosan yellow-throated marten (*Martes flavigula chrysospila*), Formosan gem-faced civet (*Paguma larvata taivana*), and other valuable wild animals. Most of these animals disappeared along with the old growth forests that were their habitats. Even during the early postwar years there were sightings of the endemic leopard cat (*Felis bengalensis chinensis*) and clouded leopard (*Neofelis nebulosa brachyurus*) in the warm-temperate mountain rain forests, and considerable numbers remained, but because tiger bones and other wild animal parts have been used as ingredients in Chinese remedies for thousands of years, hunting of the remaining leopard cats led to their extinction.

Since promulgation of the Wild Animal Breeding Law on June 23, 1989, the capture and raising of wild animals have decreased, but during the two years of 1992 and 1993 Taiwan was severely criticized by international wild animal protection organizations for having used many tiger bones and rhinoceros horns in medicines. In a bid to deflect this criticism the legal authorities beefed up control under the Wild Animal Breeding Law. Two clouded leopard skins were among the wild animals and wild animal products confiscated by Taiwanese customs authorities in 1994. Taiwan faces a major challenge in protecting its small numbers of remaining wild animals.

2-6 Energy

Energy demand has climbed dramatically with rapid urbanization and industrialization. The GDP of energy-related industries soared from NT$300 billion in 1960 to NT$4.7 trillion in 1993, for an approximate 15-fold increase. This growth comes from not only industrial demand, but also from the increased ownership of consumer appliances and automobiles that rising incomes made possible. But because nearly all Taiwan's energy has been supplied by a state-run monopoly, the sensibility and efficiency of energy policy have been repeatedly called into question, and this system has also been a major source of pollution.

Until the beginning of the 1960s most electricity was hydropower, but the long lead time for building a hydroelectric plant kept supply from keeping up with demand created by rapid economic growth, which brought thermal power to the fore. Beginning in 1965 thermal power's share of total supply surged, and for a time it accounted for 78% of the total.

In 1977 Taiwan's first nuclear power plant began operation in the north, and currently there are six nuclear reactors in operation, two each at three plants. Nuclear's share of total supply grew from 9% in 1977 to 23% in 1993, making it more important to Taiwan. As this chapter will discuss below, the problem of nuclear waste disposal will necessarily become a serious one.

Generally people who live near power plant construction sites tend to shun the plants owing to their heavy impact on the surrounding environment. In a bid to site power plants and obtain land for them, and especially to solve the siting problem of the "fourth nuclear plant"[2] and thermal plants, Taiwan in the 1980s established a system similar to that made up of the "electric power law triad" passed by Japan in 1973. This system's economic inducements made power plant siting easier for a time, but due to the heightened awareness of environmental problems that came with the groundswell in self-help and other social movements, siting opposition campaigns persisted in many instances despite economic incentives, and the system is now seen as having come to a dead end.

3. Self-Help and Pollution Disputes

Rapid industrialization and development that proceeded without consideration for the environment caused a variety of problems. But under martial law the people of Taiwan were almost totally deprived of the right to complain even when pollution damaged their health and personal assets.

---

[2] Although citizen opposition to the fourth nuclear plant continues, the project has been approved by the Legislative Yuan and the plant is under construction.

What is more, local governments often assumed the status of subcontractors to the central government, which gave top priority to industrial policy for economic growth, a situation that mostly prevented local governments from discharging their environmental management duties. Against this backdrop, the anger and discontent over pollution that worsened day by day throughout Taiwan especially from the latter half of the 1970s was increasingly expressed as "self-help" actions directed at polluters. It is important to note, however, that despite the variation in people's understanding and interpretations concerning the nature of self-help, at one time there was a tendency to use the label "self-help" for every protest action with a political or economic purpose. Here we shall avoid such confusion by narrowly defining it as environment-related actions. The "pollution dispute" concept that appeared in the latter half of the 1980s is basically an extension of self-help, but because it is specifically defined in the Pollution Dispute Management Law, its meaning is more distinct than that of self-help. The "pollution dispute" concept has been important especially for administrative work since August 1987 when the Environmental Protection Administration was created.

3-1 Self-Help

• What Is Self-Help?

The environmental departments of Taiwan's central and local governments had very poor administrative efficiency in the past, and meager ability and experience to deal with environmental problems, but this was not because there were no such problems that should have been covered by public policy. Rather, it was because the government had neglected to make environmental problems subject to public policy, which was likely because priority had been assigned to development policy that was intended to play catch-up with developed countries. Victims of pollution and environmental damage therefore abandoned hope of assistance from public institutions, turning instead to demonstrating their displeasure directly to those responsible, or carrying out protest actions demanding compensation. This is self-help, the opposite of help under civil rights. Another quality of self-help is that it protests against and tries to break with the old order, which restricted democracy. In the late stage of martial law, self-help included extraordinary actions seeking to throw off those shackles and bring about democracy and social justice. H. H. Michael Hsiao and other Taiwanese authors summarize the reasons for self-help actions in the following manner.

First, as pollution attained a level that was beyond the endurance of the citizens victimized by it, the situation objectively gave them no choice but to act. Second, polluters and victims had differing perceptions about the problem's seriousness, and therefore differing views on appropriate remedial actions. Overall, polluters had too little will to remedy the situation. Further, victims' rights were continuously violated in various ways, such as the continuance of

polluting acts while ignoring victims' protests, and failure to gain their consent. Third, public institutions that were supposed to deal with environmental problems were unable to intervene immediately because legislation giving them authority to act was belated, making them incapable of safeguarding civil rights. This elicited dissatisfaction toward the government from both polluters and victims, making citizens in particular resort to self-help instead of relying on civil rights to solve their problems.

As the foregoing analysis shows, the distortions of industrial policy that conferred top priority on economic growth were emerging, while at the same time democratic rights continued to be limited. Indeed, the feeling of distrust toward the public power responsible for this situation was the true driving force behind self-help. In other words, self-help was a movement that arose because of grave environmental damage, and in quest of democratic rights. Hsiao, et al. compiled 108 instances of self-help occurring in the 1980s, but other authors report that in the preceding decade as well there were 49 instances concerning air pollution by steel plants and cement factories alone.

• Self-Help Examples

Categorizing the 108 self-help incidents occurring from 1980 to 1987 shows that actions aimed at the chemical industry were most numerous, followed in descending order by steel and other metals, pesticide manufacturing, and municipal waste disposal, all of which account for over 10% each. Self-help actions are obviously concentrated on the heavily polluting industries. By management type, businesses targeted by self-help actions were 75.9% private and 13.0% state-run, with the remaining 11.1% being foreign-capital or other (Fig. 3). But these figures do not necessarily mean that state-run companies were less polluting or damaging to the environment. In fact, the relatively low figure for national enterprises perhaps is an indication of the difficulty that citizens had in conducting self-help actions against them, because under martial law any protest against a state-run company was a challenge toward national authority and thus dissident activity. For this reason citizens in some cases shifted part of their discontent to private-sector firms enjoying the support of national industrial policy to indirectly register their distrust of the system.

Examples of self-help incidents compiled by Hsiao, et al. show that the actions cover long periods of time (Table

Fig. 3 Ownership of Companies Subject to Self-Help Actions

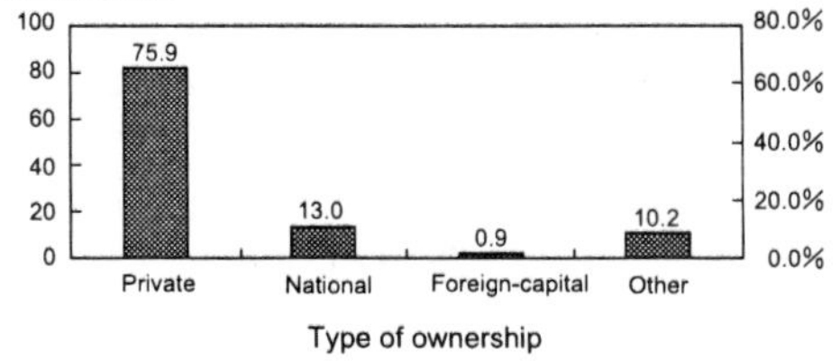

Note: Figures above bars are percentages.
Source: Hsiao, H. H. Michael, et al. *Analysis of Structure and Course of Anti-Pollution Self-Help in the 1970s*, 1988 (in Chinese).

116

4). Especially for examples (1) and (2), which had already occurred before the self-help actions, initial negotiations between the polluters and citizens failed to produce agreement owing to the three reasons described above, which induced long-term citizen protests. Both the Taichung County Tali Village Sanhuang case and the Hsinchu City Lichangjung Chemical Plant case ultimately resulted in orders to close the plants. In major investment and development projects including the Changhua County Lukang Du Pont facility, the Kaohsiung City Nantzu District Fifth Naphtha Plant, the Ilan County Sixth Naphtha Plant,[3] and the Taipei County Kungliao fourth nuclear plant, they were ultimately cancelled because of vehement citizen opposition. Those who promoted these new projects claimed that infallible measures were taken to protect the environment and that no pollution whatsoever would occur, but given the many instances in which pollution resulted, they were hardly able to gain the citizens' trust. The second nuclear waste disposal site was temporarily suspended, but subsequently there were continued shipments of wastes to Lanyu in Taitung County (an island in the Pacific), and it was impossible to contain the local citizens' anxiety. There is a close connection between this nuclear waste disposal issue and the aborted plan to ship Taiwan's nuclear wastes to North Korea for disposal. In any event, the spontaneous citizen self-help movement in the 1980s realized a number of considerable achievements.

---

[3] Abbreviated names for the "Kaohsiung Fifth Naphtha Refinery" (China Petroleum Co., Ltd.) and the "Ilan Sixth Naphtha Refinery" (Formosa Plastic Co., Ltd.). The latter plant was originally sited in Ilan, but the site was moved to Yunlin owing to citizen opposition. Thus the present Sixth Naphtha Refinery in Yunlin, which is completed and in test operation, is not the facility targeted by the self-help opposition movement.

• **Central Government Failures, Local Government Weaknesses**

We have seen that industrialization policy in Taiwan's postwar economic development had the quality of a so-called development dictatorship, and that it engendered heavy environmental impacts, but the citizens who endured harm from long-term pollution and lived with the concomitant anxiety were able to call attention to the problem only through self-help actions. Governments and other public institutions are supposed to intervene with solutions when market failures cause problems, but in reality government intervention's efficiency and possibility of success are not assured. In fact, when public policy does not see much value in the environment, governments do not try to remedy the situation, or their intervention actually worsens the damage, which is evident from typical cases such as Minamata disease in Japan. Instead of generating citizen faith in public institutions, such incidents actually result in great disappointment and antipathy toward them. Taiwan's self-help phenomenon around the 1980s was indeed a typical example of that.

One other problem that bears mentioning is that of local government in Taiwan. During Japan's period of worst pollution, many local governments showed themselves quicker and more creative than the central government by entering into pollution control agreements, beefing up ordinances and regulations, and dealing with pollution in a variety of other ways. Their achievements ultimately led to systematization of environmental policy at the national level. At that time it was local governments that served as the driving force for environmental policy. But judging by the allocation of tax and fiscal resources, and by the extent of autonomy, Taiwan's local governments are much more like small, local government agencies that carry out central government policies than autonomous units that endeavor to raise the standard of living for their citizens. Taiwan's central government is generally considered a growth-oriented development dictatorship, and it invests policy for develop-

Table 4 Major Self-Help Events of the 1980s

Case name	Location	Plant type	Period of time	Pollution type	Outcome
(1) Sanhuang Pesticide Plant	Tali Village, Taichung County	Pesticide plant	Aug 4, 1981 to Jul 31, 1986	Air, water pollution	Post-pollution settlement
(2) Lichungjung Chemical Plant	Hsinchu City	Chemical plant	Jun 3, 1982 to Apr 26, 1988	Air, water pollution	Post-pollution settlement
(3) Du Pont Chemical Plant	Lukang Village, Changhua County	Chemical plant	Jan 12, 1986 to Mar 12, 1987		Prevented
(4) Fifth Naphtha Plant	Nantzu District, Kaohsiung City	Petrochemical	Jul 2, 1987 to Apr 30, 1988		Prevented
(5) Sixth Naphtha Plant	Ilang County	Petrochemical	Oct 6, 1986 to Feb 1, 1988		Prevented
(6) Fourth Nuclear Plant	Kungliao Village, Taipei County	Nuclear power	Apr 1980 to Mar 27, 1988		Prevented
(7) Second Nuclear Waste Disposal Site	Lanyu Village, Taitung County	Nuclear waste disposal site	May 11, 1974 to Apr 25, 1988	Radioactivity contamination	Post-pollution settlement

Note: Period of time is from beginning of self-help action until settlement is reached.
Source: Based on Hsiao, H. H. Michael, et al. *Analysis of Structure and Course of Anti-Pollution Self-Help in the 1970s*, 1988 (in Chinese).

ment and economic growth with absolute top priority. It is clear from this what role is assigned to local governments. Given these circumstances of limited regional autonomy, one cannot expect Taiwan's local governments to deal with pollution as Japan's local governments have, and that translates into little progress for environmental policy.

3-2 Pollution Disputes

In addition to generating concern for the environment, self-help actions in Taiwan around the 1980s had great significance as a movement to recover and secure human rights and democracy, but still more, through the experience of dealing with this series of incidents Taiwan's government agencies internalized and institutionalized ways of addressing environmental problems. Especially since creation of the Environmental Protection Administration in August 1987, administrative management for controlling pollution, dealing with pollution problems, environmental education, and other environmental concerns has progressed to the extent one might expect. This section will, along the lines of self-help, analyze how the government has legally dealt with issues that it has defined as pollution disputes.

• Legal Definition of Pollution Disputes

Article 2.2 of the Pollution Dispute Management Law promulgated on February 1, 1992 defines pollution as anthropocentric causes that damage the environment where people live, and that harm or may harm the citizens' health. The range of pollution is defined as water pollution, air pollution, soil contamination, noise, vibration, offensive odors, wastes, contamination by toxic materials, ground subsidence, contamination by radioactivity, and any other type designated by central authorities. Pollution disputes are therefore defined as disputes caused by, or that could be caused by, loss or damage due to the above 10 kinds of pollution and others designated by the central authorities. Thus, while essentially pollution disputes and self-help are more or less the same, they differ in that pollution disputes are legally defined and have prescribed procedures for settling disputes, while self-help does not.

The *Report on Pollution Dispute Settlements* was published in 1992 and updated in 1993. The 1992 edition mainly

gave the definition of pollution, the principles for dealing with it, and other legal grounds, related how pollution disputes occurring over the three years from 1988 to 1990 had been managed, and described the future challenges of environmental management agencies. To increase the store of experience in dealing with disputes, the 1993 edition included analyses of instances up to December 1992, and considerably enhanced its material. Having analyzed 259 cases, the report goes on to write about the causes of pollution disputes, stating, "The causes of most pollution disputes are not due to improper motives of the protestors or attempts at profiting themselves, but rather because of flaws in the siting processes used when factories were originally established, and because responsibilities for administrative management were not properly discharged. The reasons for the latter problem include inadequate legal controls, insufficient personnel and knowledge in the implementing agencies, and polluters' lack of a sincere desire to solve problems." This closely coincides with analyses on why self-help incidents happen, making it highly significant that administrative management agencies are admitting their own responsibility.

• Pollution Dispute Analysis

Analyzing the 259 cases recorded in the 1993 edition of *Report on Pollution Dispute Settlements* shows that 108 of them are instances of self-help actions, which in particular include all seven cases in Table 4.

Plotting the 259 cases according to time of occurrence (Fig. 4) reveals an explosive increase in pollution disputes during the approximately one year from September 1988 to September 1989. September 1988 marked the occurrence of the Linyuan incident, a major accident that polluted inland and ocean waters over a broad area when an underground oil distribution pipe at a China Petroleum Company plant in Kaohsiung County Linyuan Village burst. Victims of the accident were not limited to nearby residents, for the widespread marine pollution was also a considerable blow to regional marine aquaculture. China Petroleum ultimately resolved the matter by paying a total of NT$1.27 billion in reparations to victims through their towns and villages. Because this set a precedent, pollution victims began demanding compensation according to the damage caused.

Fig. 4 Pollution Disputes by Time of Occurrence (Jan. 1988 to Sept. 1992)

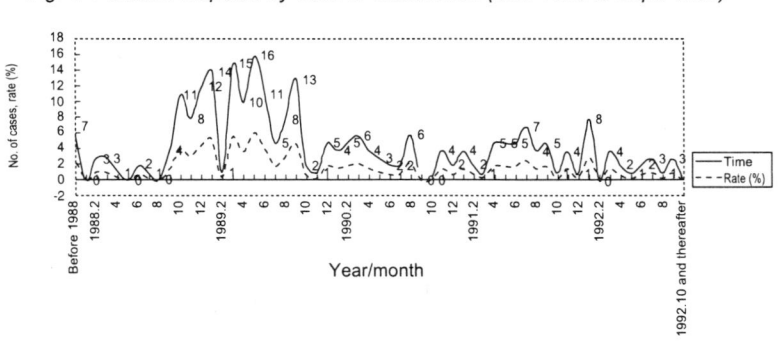

Source: Prepared from 1993 edition of *Report on Pollution Dispute Settlements.*

This incident had considerable impact because the dispute involved almost all the large petrochemical and chemical plants in the Linyuan Industrial Zone or nearby petrochemical complexes. Based on experience in settling pollution disputes during the period in which many of them occurred, administrative authorities and even private companies are now to a certain degree able to determine the specifics, background, and extent of damage in pollution incidents, leading to the establishment of standards for seeking compensation and guidelines for settling disputes. Further, polluters started dealing more actively with their pollution, which contributed to a substantial reduction in severe clashes. The low number of eight incidents in January 1992 marked the beginning of a trend toward a low occurrence rate. Another fact deserving note is that owing to heavy opposition and protests in the Kaohsiung City Nantzu District Fifth Naphtha Plant case listed in Table 4, opposition registered in a May 6, 1990 referendum led the developer, China Petroleum, to donate a local construction fund of NT$1.5 billion to the local government of the Nantzu District's Houchin area. It remains to be seen what significance and problems this kind of solution will present.

Pollution disputes are concentrated in southern areas such as Kaohsiung County and Kaohsiung City, as well as Taoyuan County, which host heavily polluting industries like heavy manufacturing, chemicals, oil products, and steel, but many small chemical plants and metal plants, as well as nuclear power plants, are scattered throughout the northern region as well, including Taipei and Miaoli counties, whose dispute occurrence rates are not low.

An examination by industry type of companies causing pollution disputes (Fig. 5) shows that the chemical industry, which was most often targeted in self-help actions from the latter half of the 1970s to the first half of the 1980s, accounted for a smaller proportion of the whole, perhaps because some companies eagerly made effective improvements. On the other hand, relatively large petrochemical companies like China Petroleum caused one serious accident after another owing to aging equipment, operation errors, inadequate monitoring, and other causes. This produced more victims over larger areas and increased the number of disputes. Although by nature steel and other metals are large-scale, and pesticide manufacturing is highly haz-

ardous and polluting, small steel and metal plants were forced by stiff market competition into merging or shifting operations to developing countries in Southeast Asia and elsewhere, which decreased the number of individual disputes. Pesticide manufacturing finds demand for chemicals is changing[4] due to the transformation of the industrial structure. A number of polluting plants including the one at Taichung County Tali Village, Sanhuang were closed under orders to shut down, which mitigated air pollution.

Waste disposal increased from 10.2% to 17.4%, which has brought to light problems related to the disposal of municipal solid wastes and industrial wastes, such as groundwater contamination and odors from waste landfills, and the siting of new landfills and incinerators. There is also a trend toward increasing disputes relating to electric power, such as thermal power plant air pollution and nuclear power plant siting (the Taipei County Kungliao fourth nuclear plant).

A comparison of pollution sources causing pollution disputes (Fig. 6) and self-help actions (Fig. 3) by form of management shows that in pollution disputes private companies fall from 75.9% to 44.8% while national companies increase substantially from 13.9% to 49.8%, which probably reflects a decrease, due to the lifting of martial law, in the institutional difficulty of criticizing and mounting opposition movements against environmental damage caused by state-run companies. The fact that it became possible for pollution victims to protest directly against national enterprises and to demand remedial measures and compensation can be considered a sign of democratic progress in Taiwanese society, and it also indicates that pollution by national companies is especially grave. China Petroleum in particular is constantly coming up in pollution disputes because of its reluctance to alleviate pollution, despite its enjoyment of business profits that are high even for a state-run company.

A study to determine who lodges protests in pollution disputes shows that it is victimized citizens in 74.6% of the cases, environmental organizations in 7.4%, and local governments in 2.8%. Even though citizens are not organized, they account for the overwhelmingly large proportion of protest actions. Increasing disputes over factories, waste disposal sites, and the siting of incinerators, electric power facilities and the like show that citizens fear pollution and are actively protesting.

Fig. 5 Dispute Pollution Sources by Industry

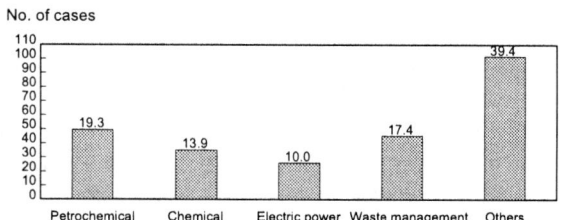

Note: Figures above bars are percentages.
Source: Prepared from 1993 edition of *Report on Pollution Dispute Settlements.*

Fig. 6 Dispute Pollution Sources by Management Type

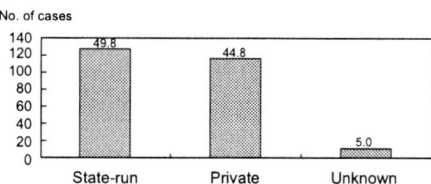

Note: Figures above bars are percentages.
Source: Prepared from 1993 edition of *Report on Pollution Dispute Settlements.*

---

[4] Overall pesticide use is decreasing, although the need for quick harvests increases pesticide inputs per unit land area.

4. Summation

Taiwan pursued rapid industrialization under martial law, a time when administrative authority was supreme, and everything including legislation and the judicature was integrated under a governing system whose fundamental principle was catch-up development. As consideration for the environment was hardly accorded its rightful place, environmental damage caused widespread harm, but the citizen reaction to this coincided with nascent efforts for democratization, and intensified the self-help movement. Taiwan's environmental policy, which had its real beginning with the 1987 establishment of the Environmental Protection Administration, was in part meant to address these democratization and self-help movements, and that was why policy legally defined confrontations over environmental damage as pollution disputes, and institutionalized procedures to deal with them. In this respect, Taiwan's environmental policy began as a way of dealing with environmental problems and their attendant disputes after they had arisen.

Subsequently there was progress in local autonomy, such as the holding of mayoral elections in the cities of Taipei and Kaohsiung, and there is a trend toward the democratization of institutions. The question is whether the system of environmental laws and policies created in this manner can effectively control the causes of environmental damage. We must watch to see what kind of new challenges arise as the world economy is globalized, and how they will force changes in that system.

(CHEN Li-chun, UETA Kazuhiro)

Part III   Indicators

[1] Basic Economic Indicators: External Debt and International Trade

The debt crisis of developing countries became apparent in the early 1980s mainly in Central and South America. A look at the net flow of funds around the world — payments on interest and principal subtracted from newly introduced funds — shows that beginning in the mid-1980s money tended to flow back from the developing to the developed world (Table 1).

Even some Asian countries have external debt problems despite the region's rapid economic growth. Asian countries obtain a great deal of funding especially from Japan. The extent of a country's external debt is generally indicated by its debt service ratio (DSR), which is the ratio of annual debt repayments (payments on interest and principal) to the monetary amount of exports, the reason being that basically external debt must be repaid with foreign currency earned with exports. It is often said that a debt crisis emerges when the DSR exceeds 20%. In 1993 some Asian countries had very high figures, such as 24.9% for the Philippines, 28.0% for India, and 31.8% for Indonesia. One cause is the large amounts of yen credit from Japan. The rapid appreciation of the yen in the latter half of the 1980s increased direct investment from Japan in other Asian countries, but it was also a major factor that increased their debt burdens. Especially indicative of this are the high figures for the Philippines and Indonesia, which accepted especially large amounts of yen credit from Japan. China too has many long-term debts whose repayment term has not yet arrived, and these could turn into a serious problem in the future.

Until now trade has been a driving force behind Asia's economic growth. Since the 1970s the NIEs of South Korea and Taiwan have been in the world spotlight for their rapid growth. They pursued export-oriented industrialization, which contrasts with the Central and South American NIEs that adopted a strategy of industrialization to substitute for imports, a strategy that later quickly lost momentum. A characteristic of Asian NIEs was how they took advantage of their cheap labor to import intermediate goods from developed countries, then process and export them. Continually increasing exports naturally led to increasing imports (Fig. 1). This strategy was in time taken over also by ASEAN countries. Beginning in the 1980s a number of Asian countries racked up double-digit annual increases in trade and achieved astonishing growth about twice the rates of developed countries (exports 5.1%, imports 5.8%) during those same years. Another fact worth noting is that although trade until then was based on a Japan-U.S. axis, in recent years an

Table 1 Net Transfer from Developed to Developing Countries

(billions of US dollars)

1970	4.46
1980	30.23
1984	-9.41
1985	-17.64
1986	-21.01
1987	-29.69
1988	-31.49
1989	-25.17
1990	-21.62

Source: World Bank, *World Debt Tables 1991-92.*

Fig.1 Direct Investment Balance among Japan, the US, the EU, and East Asia

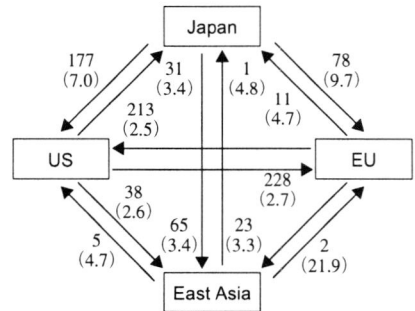

Note: Figures next to arrows are 1993 investment balances (billions of US dollars), and parenthetical figures are multiples of 1985 balances.

Data: OECD, *IDIS*; IMF, *IFS*

Source: Ministry of International Trade and Industry, ed., 1996 *Trade and Commerce Report, Overview*, p. 136.

increasing proportion of trade is occurring among Asian NIEs and ASEAN countries. It is clear why Asia is called the center of world economic growth.

Recently the world economy is becoming globalized and borderless, and throughout the world countries are promoting economic deregulation and free trade. Developing countries too are being swept along by this tide. One factor that has induced them to adopt such policies is the "structural adjustments" involved in the debt problem. Since the 1980s the International Monetary Fund and the World Bank have demanded as a loan conditionality for heavily indebted countries that they set up arrangements making it possible to repay their loans, and for that purpose the IMF and the

World Bank have pressured those countries to restructure their economies. These "structural adjustment policies" include requirements such as austerity budgets, fiscal reorganization through tax reform and other measures, privatization of state-run and public companies, liberalization of trade and foreign exchange, deregulation, and subsidy reductions. The aim of this neoclassical school thinking is small government, and an attempt at a thoroughgoing review of the inefficient economic management and policy followed heretofore. This structural adjustment policy prompted the opening of economies and trade liberalization not only in Asian countries, but also throughout the developing world. Some countries realized improvements in their macroeconomic indicators, but there is also much criticism that owing to subsidy reductions, decreased public services in line with the principle that the beneficiary should pay, and other changes, structural adjustment has had grave impacts on society's weak.

Another fact we should not forget in considering "liberalization" in Asia is that beginning in the late 1970s socialist countries like China and Vietnam adopted policies for reforming and opening their countries to the outside. Especially the rapid economic growth of China with its huge population of 1.3 billion will likely have a heavy impact on not only Asia, but the whole world economy.

(OTA Kazuhiro)

[2] Basic Economic Indicators: Official Development Assistance

Development assistance by developed countries tends to increase on the whole (Fig. 1). Japan's Official development assistance (ODA) in particular has increased rapidly, overtaking the U.S. in 1989 and making Japan the world's number one donor country, a position it maintains to this day.

While all donor countries have increased their development assistance, not many OECD countries have attained the goal of 0.7% of GDP set by the Development Assistance Committee (DAC), and that is the reason that, at international venues like the Earth Summit, developing countries repeatedly press the developed countries to increase their assistance (Table 1). In the 1990s developed countries started feeling the burden of their assistance, and in an increasing number of instances the donors have made cuts. Some of the reasons for these reductions are the end of the Cold War, which was one motive for giving aid, developed countries' economic downturns and fiscal difficulties, and the fact that results are not always apparent despite long years of assistance.

Table 1 1995 ODA by Primary DAC Aid Donors

Ranking	Country	ODA(millions of US dollars)	Share (%)	Percent of GNP (%)
1	Japan	14,489	24.5	0.28
2	France	8,439	14.3	0.55
3	Germany	7,481	12.6	0.31
4	US	7,303	12.3	0.1
5	Netherlands	3,321	5.6	0.8
6	Britain	3,185	5.4	0.29
7	Canada	2,311	3.9	0.42
8	Sweden	1,982	3.2	0.89
	DAC totals	59,205	100.0	0.27

Source: Ministry of Foreign Affairs, ed., *Japan's Official Development Assistance*, 1996, vol.1.

Specific elements of development assistance are characterized by the donor countries. For example, a high proportion of aid from northern European countries is grants, but much of that is tied aid. Assistance by the U.S. originally had a strong strategic orientation because of the Cold War, and even in the 1990s there has been no great change in that tendency. Characteristics of Japanese assistance include: 1) a high proportion of credit, 2) a big allocation to

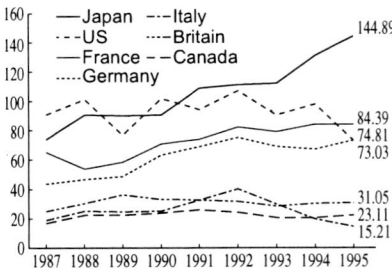

Fig.1 ODA by Primary DAC Countries
(net disbursement base, millions of US dollars)

Source: Ministry of Foreign Affairs, ed., *Japan's Official Development Assistance, 1996*, vol. 1, p. 23 (In Japanese). Based on a 1996 DAC press release.

large-scale economic infrastructure, and 3) a quickly falling proportion of tied aid.

To developing Asian countries Japan has a big presence as a donor country. As of 1994, Japan accounted for 51.8% of the ODA in Asia (Fig. 2). Because ODA accounts for large percentages of the national budgets of low-income countries like Bangladesh, assistance is of major significance to them.

In recent years some Asian countries have been giving assistance to other countries while receiving aid themselves, but there is considerable variety in the aid they provide, such as that given for political reasons, or because the donors themselves are enjoying rapid economic growth. South

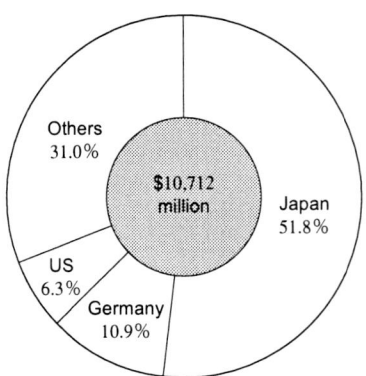

Fig.2 Aid in Asian Region by Primary Donors
(1994, net disbursement base)

Source: Ministry of Foreign Affairs, ed., *Japan's Official Development Assistance 1996*, vol. 1, p. 28.

Korea and Taiwan in particular will henceforth make their presence as donor countries increasingly felt.

A frequent observation is that many aid projects have severe environmental impacts because they put such a premium on economic benefit. One manifestation of this would be what is called "pollution export," in which businesses in developed countries transfer the polluting processes of their manufacturing and other industries to the developing world to avoid increasingly stringent environmental regulations at home. Typical examples of this are ODA involvement in the establishment of a steel manufacturing and sintering plant and a copper refinery in the Philippines in the 1970s.

One concern, owing to increasingly strict standards for aid project implementation, is the heavy impacts of projects on the environment and area inhabitants. Typical examples of this are construction of the Kotopanjang Dam in Indonesia, and dam construction on India's Narmada River. The latter is the first instance in which a project came under reconsideration because of a growing campaign over concern for impacts on the environment and inhabitants, this despite the funding decisions by governments and international agencies including the World Bank.

Since the latter half of the 1980s development assistance has been pressed to take environmental concerns into account because of more vocal world public opinion and the international environmental conferences growing out of that public outcry. Since the DAC proposed in 1985 that aid projects have environmental impact assessments, it has issued a number of environment-related recommendations, and aid by a number of countries has gradually incorporated consideration for the environment. In Japan, environment-related items are now considered in preliminary studies, and some projects are themselves implemented for the purpose of environmental conservation. Some examples of the latter are the establishment of "Environment Centers" in Thailand, Indonesia, China, and other countries, the Gujarat Forestry Development Project in India, and the Environmental Infrastructure Support Credit Program in the Philippines. Nevertheless, such environment-related ODA has just begun, so it remains to be seen how effective it will be.

(OTA Kazuhiro)

Table 2 Basic Economic Indicators

		1 China	2 Indonesia	3 Japan	4 South Korea	5 Malaysia	6 Philippines	7 Thailand	8 Taiwan
A.	Gross domestic product (GDP)								
A.1	GDP (millions of dollars, 1993)	425,611	144,707	4,214,204	330,831	64,450	54,068	124,862	226,243
A.2	GDP growth rate (%, 1980-93)	9.6	5.8	4.0	9.1	6.2	1.4	8.2	10.0
A.3	Per capita GDP (dollars, 1993)	490	740	31,490	7,660	3,224	850	2,110	10,852
A.4	GDP by industry type (%, 1993)								
	Primary	19	19	2	7	29	22	10	4
	Secondary	48	39	41	43	25	33	39	36
	Tertiary	33	42	57	50	46	45	51	60
B.	Trade								
B.1	Exports (millions of dollars, 1993)	91,744	33,612	360,911	82,236	47,122	11,089	36,800	84,678
B.2	Avg. export growth rate (%, 1980-93)	11.5	6.7	4.2	12.3	12.6	3.4	15.5	10.0
B.3	Imports (millions of dollars, 1993)	103,088	28,086	240,670	83,800	45,657	18,757	46,058	77,099
B.4	Avg. import growth rate (%, 1980-93)	4.2	4.5	6.3	11.4	9.7	4.5	13.8	13.2
B.5	Current account balance (millions of dollars, 1993)	-11,344	5,526	120,241	-1,564	1,465	-7,668	-9,258	7,579
C.	Finance								
C.1	Annual gov't revenues (percent of GDP) (%, 1993)	5.2	19.4	15.4	18.9	28.7	17.1	18.3	-
C.2	Fiscal surplus/deficit (percent of GDP) (%, 1993)	-2.3	0.7	-1.6	0.6	1.7	-1.5	2.1	-
D.	Official development assistance (ODA)								
D.1	ODA accepted (millions of dollars, 1985)	772	603	-3.797	-9	229	486	481	-
	ODA accepted (millions of dollars, 1991)	2,286	1,783	-10,952	-134	459	1,231	738	-
D.2	Foreign debt (millions of dollars, 1993)	83,800	89,539	-2,180,880	47,203	23,335	35,269	49,819	-89,039
D.3	Debt Service Ratio (%, 1993)	11.1	31.8	-	9.2	7.9	24.9	18.2	-
D.4	ODA portion of annual gov't revenues (%, 1993)	7.7	45.4	-	-	0.7	16.4	57.9	-
D.5	Gold and foreign currency reserves (millions of dollars, 1993)	21,199	12,355	95,589	20,262	28,294	5,921	25,439	89,290

Sources: A3-5 and Taiwan data are from *Annual Report on Asian Economic Trends 1995* by the Institute of Developing Economies; other data are all from the World Bank's *World Development 1995*.

[3] Worker Accidents and Occupational Illnesses

Rapidly advancing industrialization in Asia brings increasing reports of occupational accidents and illnesses (Figs. 1 and 2). Governments, labor, and management have a rising awareness of this problem's seriousness, although accurate figures are hard to come by. In all countries statistics for labor accidents are kept for enterprises that employ at least a certain number of people, making it possible that injuries are not reported for small businesses and temporary employees. Further, differences among countries in the way labor accident statistics are interpreted is another reason for caution. Dividing Asian countries into the following three groups helps clarify this problem.

First are countries and regions that keep basic statistics on labor accidents, such as Japan, South Korea, Hong Kong, and Singapore. In the last three, which industrialized comparatively early, worker accidents have decreased in recent years. In the second group, including countries like Thailand and Malaysia, the recording of worker accident statistics has revealed the swift increase in such accidents and shows the urgent need for remedial measures. These countries have a particularly large number of accidents in their construction and manufacturing industries. Labor accidents among Malaysia's more than 2 million foreign workers (about half of which are there legally) have developed into a social issue. Group three countries, such as China and India, lack consistent nationwide coverage of worker accident statistics and badly need practical ways to address this problem. However, it is important to note that some accidents may go unreported in countries of all three groups, and that there are differences among countries in their positions on labor accident reporting.

Incidences of occupational illnesses, meanwhile, are conceivably far higher than statistics show for several reasons, such as physicians not being accustomed to diagnosing such illnesses, and the lack of reporting systems. Apart from government statistics, these days one occasionally finds occupational illness case reports by physicians or industrial hygiene specialists. Some of the reported disorders are hardness of hearing due to noise, poisoning from chemical substances or pesticides, pneumoconiosis, byssinosis, and diseases of the musculoskeletal systems. It is also unfortunate that even since the 1984 tragedy in Bhopal major accidents that kill several to several dozen people continue to happen in Asia. Reducing the risk of serious accidents like these necessitates international joint efforts and the dissemination of suitable management methods.

This situation led to better initiatives for safety and health throughout Asia, and there were many concrete efforts from the latter half of the 1980s and into the 1990s. Noteworthy government-led efforts include establishment

Fig. 1 Occupational Injuries in Thailand and Malaysia, 1986-94

(thousands of persons)

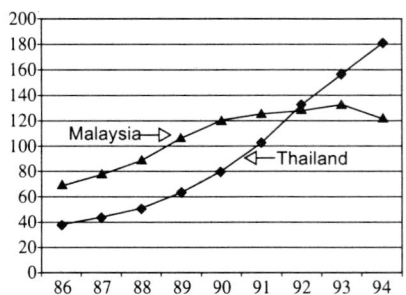

Source: International Labor Office, *Yearbook of Labor Statistics*, 1996.

Fig. 2 Fatal Occupational Injuries per Thousand Workers, 1992 (1989 data for Philippines)

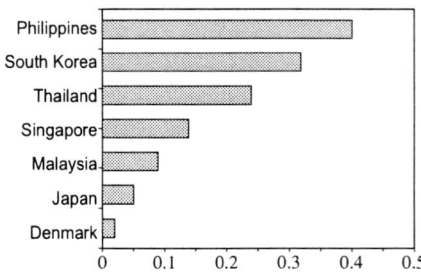

Table 1 Improvements Proposed and Implemented by Small and Medium Enterprises in the WISE Program, 1994-96

	Technical area	Proposed	Implemented
1	Materials handling and storage	409	364
2	Workstation design	167	134
3	Productive machinery safety	166	142
4	Control of hazardous substances	116	97
5	Lighting	250	225
6	Premises	482	373
7	Work-related welfare facilities	239	196
8	Work organization	185	150
9	Protection of general environment	46	43
	Totals	2060	1724

Source: DOLE/UNDP/ILO, *WISE Awareness Presentation Package 1997.*

of the Ministry of Labor and Social Welfare in Thailand and passage of the Occupational Safety and Health Act in Malaysia. In the Philippines, the government, labor, and management, with UNDP financial cooperation and ILO technical assistance, cooperated in pursuing a participatory workplace improvement project for small and medium enterprises called Work Improvement in Small Enterprises (WISE), and have realized some success (Table 1). Some of the important elements that go into workplace improvements are (1) begin with a focus on low-cost improvements that can be effected immediately on site, (2) use action check lists to identify where improvements should be made, and (3) use group work methods to encourage positive participation by each individual. Even labor union movements have begun to see the importance of improvements in non-wage profits, such as occupational safety, health, and employee welfare

facilities. Projects built around local labor unions include the POSITIVE (Participation-Oriented Safety Improvements by Trade Union Initiatives) Program implemented by the Japan International Labour Foundation, and WE-OSH (Workers' Education — Occupational Safety and Health) by DANIDA-ILO. Such programs have brought about concrete improvements in safety and health by training local safety and health trainers, and developing easy-to-use, action-oriented training packages. Common to all these labor-management efforts in Asia is a forward-looking, practical approach aimed at improvements made possible by local conditions. Approaches like this were used also in many of the successful initiatives that realized concrete achievements through international technical cooperation in the area of occupational safety and health.

(KAWAKAMI Tsuyoshi)

Table 2 Occupational Injuries and Diseases

		China	Japan	South Korea	Malaysia	Philippines	Thailand	India	Singapore
A.	Employment								
A.1	Employed workers (thousands, 1995)	614,690	64,570	20,377	7,645	25,698	32,095	27,525	1,700
A.2	Working population by industry [year surveyed]	[1993]	[1995]	[1995]	[1995]	[1995]	[1991]		[1995]
	Primary industries (thousands)	339,660	3,670	2,541	1,527	11,323	18,777	-	4
	Mining (thousands)	9,320	60	27	33	95	54.5	-	1
	Manufacturing (thousands)	92,950	14,560	4,773	1,781	2,571	3,465.1	-	408
	Electricity, gas, and water (thousands)	2,400	420	69	48	103	110.3	-	6
	Construction (thousands)	30,500	6,630	1,896	611	1,239	1,178.4	-	113
	Transportation, storage, and communication (thousands)	16,880	4,020	1,068	359	1,489	833.9	-	183
	Services (thousands)	106,750	34,950	9,799	3,287	8,855	6,702.9	-	814
	Others (thousands)	3,740	260	207	0	21	15.2	-	173
B.	Worker accidents								
B.1	Workers injured (persons, 1991)	-	200,633	128,169	124,898	50,860	103,296	276,416	5,156
	1992 (persons)	-	189,589	107,435	130,019	68,240	131,800	236,596	4,715
	1993 (persons)	18,122	181,900	90,288	133,293	73,020	156,550	183,391	4,257
	1994 (persons)	16,271	176,047	85,948	122,688	45,840	181,640	143,790	4,019
	1995 (persons)	-	167,316	78,034	114,134	-	-	103,793	3,948
B.2	Fatal injuries (persons, 1995)	7,235	2,414	2,662	828	220	820	1,412	65
B.3	Workers injured, by industory [year surveyed]	[1994]	[1992]	[1995]	[1995]	[1994]	[1994]	[1990]	[1995]
	Primary industries (persons)	191	4,640	403	20,465	1,670	0	-	0
	Mining (persons)	4,957	1,093	1,889	1,016	220	721	1,565	-
	Manufacturing (persons)	6,507	53,653	36,228	62,483	30,690	77,287	98,458	2,833
	Electricity, gas, and water (persons)	-	-	140	542	360	364	1,369	11
	Construction (persons)	2,670	54,357	22,542	4,406	2,440	15,629	7	893
	Transportation, storage, and communication (persons)	773	18,603	8,963	4,826	1,450	2,048	570	127
	Services (persons)	449	-	-	17,298	9,010	6,982	-	84
	Others (persons)	724	57,243	7,869	3,098	-	265	-	0
C.	Total occupational diseases (persons, 1992)	-	10,842	1,328	2,942	-	62	-	900

Notes: A1-China: 1994 values.
B3-South Korea: Services are included in others.
B1-India: 1986-1990 values.
Sources: ILO, *Yearbook of Labour Statistics 1996*.
ILO/EASMAT, *Occupational Accidents and Diseases — A Statistical Survey for Selected Countries in Asia and the Pacific*, 1995.
H. Ono and K. Enomoto, ed., *Profile on Occupational Safety and Health in India*, ILO/ARPLA, 1992.
K. Kogi, et al., *Low-cost Ways of Improving Working Conditions: 100 Examples from J. Thurman, Higher Productivity and a Better Place to Work*, ILO, 1988.

[4] Health and Education

Asia's rapid economic growth is in the world spotlight. On visits to Southeast Asia's major cities such as Bangkok, Kuala Lumpur, Manila, and Ho Chi Minh City one cannot help but notice two facets: socioeconomic problems including traffic congestion, air pollution, slums, and poverty on the one hand, and the smiling faces of populations brimming with vitality on the other. How can health and education be improved through the best initiatives of the local people themselves?

Statistics on the mortality rate for children under age 5 in ASEAN 10 countries with great future promise for intraregional cooperation (excluding Hong Kong and Brunei) show that the rate has dropped in all those countries over the last 35 years, although the rate of decline differs from one country to another (Fig. 1). As of 1995 there were countries like Singapore, which can boast of world-class sanitation and hygiene, and countries like Cambodia, Myanmar, and Lao PDR, where more than one in 10 children do not live to the age of 5. Countries with high child mortality rates have also been shaken by war and political instability. In 1960 the Philippines had the highest health level in its region, but subsequently improvements in health and hygiene were delayed owing to the domestic situation. Differences among countries and social classes in the prevalence of tuberculosis, gastrointestinal infections, AIDS, and other illnesses too are due to socioeconomic conditions.

Bringing education to more citizens plays a preeminent role in improving health and hygiene. Although literacy rates are generally high in Southeast Asia, East Asia, and Sri Lanka, there is much room for improvement in South Asian countries. And while literacy rates are more or less the same for men and women in the Philippines, Thailand, Vietnam, and Sri Lanka, rates among women are low in other countries (Fig. 2).

Education of children is of concern because they are the future. In some Asian countries almost all children are able to attend elementary school, but attendance rates such as 44% in Pakistan and 79% in Bangladesh show that many Asian children still cannot attend (Table 2). Furthermore, of 10 children starting elementary school, four in India, three in the Philippines, and one in China and Thailand drop out before finishing. Middle school attendance rates are lower; even in Thailand with its high economic growth rate, only half the children move up to middle school.

A crucial challenge for Asia is how to promote equal opportunity in health and education, which constitute the

Fig.1 Mortality Rates of Children under Five in Southeast Asian Countries (comparison of 1960 and 1995 mortality rates per thousand births)

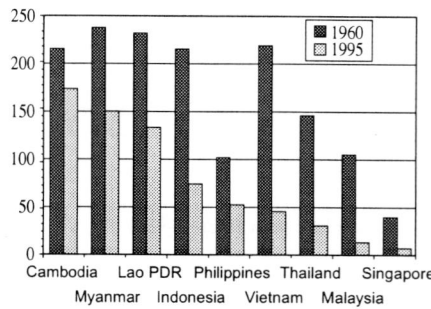

Source: UNICEF, *The State of the World's Children 1997.*

Fig.2 Comparison of Male and Female Adult Literacy Rates, 1995

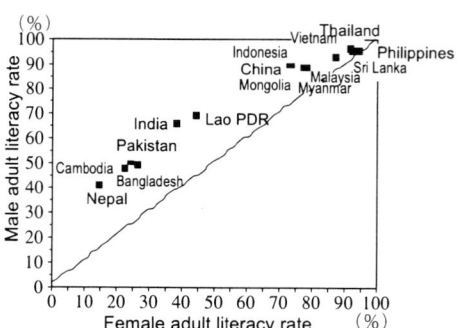

Source: UNICEF, *The State of the World's Children 1997.*

Table 1 Group Discussion Results on Improvement Needs of Living and Working Conditions Among Mekong Delta Farmers, Vietnam

Daily life improvements	Labor improvements
Hygienic meals	Wearing shoes in the fields
Sanitary eating utensils and clothing	Long sleeves for protection from sun
Sanitary lavatories	Safe storage of pesticides
Household medicines	Boats to transport farm produce
Ventilation of dwellings	Rest sites in rice paddies
Toys for children	Clean drinking water
Planned household finance	Lavatories near farms
Family discussion and cooperation	Joint purchase of agricultural machines

Source: Can Tho Department of Health, Vietnam and Institute for Science of Labour, Japan.

basis of everyday life. Further, the situation has changed from the time when poverty and sickness were equated with rural areas. Although formerly it was important to address problems of rural illness and poverty, a new and important item on the agenda is how to improve the health of urbanites, especially those with low incomes, factory workers, and country people who are in the cities to earn money.

The following items are important in bringing about improvements in health and education.

First is building the medical and educational infrastructures that allow people to share equally in the benefits of economic development. For instance, setting up health insurance systems is an important task in Southeast Asia with its fast economic growth. A lack of action would result in a double medical care system in which the affluent would use first-rate private hospitals with facilities like those in the developed countries, while the poor would use public hospitals which, although possibly having superlative staff members, are crowded and less well equipped.

Second, these countries must energetically promote specific preventive measures for tuberculosis, AIDS, and other high-risk illnesses while maintaining coordination with other sectors. International technical cooperation still has a primary role to play here.

Third, it is vital to develop and put into practice a methodology that reinforces and supports the initiatives of citizens themselves. The results of group discussions at a seminar on the self-improvement of daily life and labor in a Vietnamese Mekong Delta village (Table 1) showed that the villagers, through participatory training, were able to note the improvements needed in their own livelihood infrastructure. A follow-up investigation confirmed that villagers were indeed capable of making many improvements under their own initiatives.

(KAWAKAMI Tsuyoshi)

Table 2 Health and Education

		1 China	2 Japan	3 S. Korea	4 Malaysia	5 Philippines	6 Thailand	7 India	8 Singapore
A.	Socioeconomic indicators								
A1	Total population (millions)	1,221.5	125.1	45.0	20.1	67.6	58.8	935.7	2.8
A2	Per capita GNP (1994, US$)	530	34,630	8,260	3,480	950	2,410	320	22,500
A3	Household income distribution (%, 1990-94)								
	Lowest 40%	17	22	20	13	17	14	21	15
	Highest 40%	40	38	42	54	48	53	43	49
B.	Health and population indicators								
B1	Average life expectancy at birth, males (1995)	68.2	76.8	68.8	69.9	66.6	65.2	62.6	73.5
	Average life expectancy at birth, females (1995)	71.7	82.9	76.1	74.3	70.2	71.6	62.9	78.6
B2	Annual no. of births (thousands, 1995)	21,726	1,278	736	543	1,975	1,124	26,106	43
B3	Under 5 mortality rate (per 1,000 births, 1995)	47	6	9	13	53	32	115	6
B4	Infant (under 1) mortality rate (per 1,000 births, 1995)	38	4	8	11	40	27	76	5
B5	Maternal mortality ratio (per 100,000 births, 1990)	95	18	130	80	280	200	570	10
C.	Nutritional indicators								
C1	Sufficiency of per capita necessary daily calories (%, 1988-90)	112	125	120	120	104	103	101	136
C2	Underweight child birth rate (%, 1990-94)	9	7	9	8	15	13	33	7
D.	Educational indicators								
D1	Total primary school enrollment, males (%, 1990-94)	120	102	97	93	108	98	113	109
	Total primary school enrollment, females (%, 1990-94)	116	102	99	93	107	97	91	107
D2	Percentage of cohort reaching grade 5 (%, 1990-94)	88	100	100	98	67	88	62	100
D3	Total secondary school enrollment, males (%, 1990-94)	60	95	97	56	64	38	59	69
	Total secondary school enrollment, females (%, 1990-94)	51	97	96	61	65	37	38	71

Note: Total attendance rate in D1 and D3 is the ratio of children attending school, regardless of school age, to total school-age population.

Sources: UNICEF, *The State of the World's Children 1997*.UNDP, *Gender and Human Development: Human Development Report 1995*.

[5] Trends in Agriculture and Food Production

The past few years have seen heated discussion over the future of world agriculture and food production. This controversy was triggered by the Worldwatch Institute's Lester Brown with his predictions in *Full House*, and his sensational book *Who Will Feed China?* Grounds for his contention are the limitations imposed by shrinking agricultural resources, such as land degradation and water shortages, and the resulting effects on rising population and increasing food demand.[1]

For example, worldwide per capita arable land clearly diminished from 0.44 ha in 1961 to 0.26 ha in 1994 (Fig. 1). Especially in East Asia with its heavily populated countries, per capita arable land fell from 0.18 ha to 0.10 ha within the same timeframe.[2]

Of course as the FAO points out, the land/population ratio is only one of the many factors determining the per capita food supply.[3] While East Asia's land/population ratio worsened during the last three decades, the region has at the same time made remarkable improvement in its food supply. All seven major countries increased their per capita daily caloric supply, and in 1994 all of them crossed the 2,300-calorie mark.

But while per capita grain production increased in China, Indonesia, and the Philippines, it fell in Japan, South Korea, Malaysia, and Thailand (Table 1). Overall East Asian grain imports grew quickly from 11.3 million tons in 1961 to 75.8 million tons in 1995, leaving Thailand as the only net grain exporter (Fig. 2, Table 1). The reason that this decline in domestic agricultural production did not translate into food problems is that these countries' economic growth increased their ability to import food.

The FAO's perception — that if food were equally distributed, everyone on the planet would have enough to eat — may be sound only under present circumstances.[4] But it is a moot question whether in the future it will still be possible to argue that the problem is not a food shortage, but poverty (i.e., lacking the right to acquire food).

First of all, in addition to the conflict with non-agricultural uses of the land and the water resources suited to agricultural production, those resources are also damaged by

Fig.1 Per Capita Arable Land and Grain Producing Area Worldwide and in East Asia, 1961-95

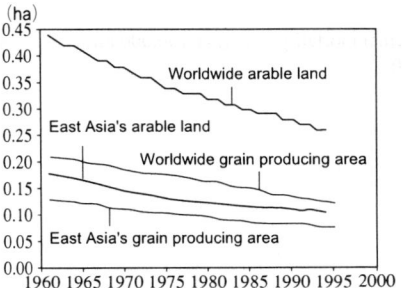

Source: FAO, *FAOSTAT Agricultural Data*, 1996.

Fig.2 East Asia's Net Grain Trade, 1961-95

(millions of tons)

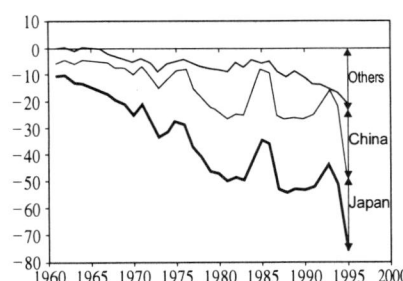

Source: FAO, *FAOSTAT Agricultural Data*, 1996.

Fig.3 Meat Production and Animal Feed Proportion of Domestic Grain Demand in East Asia, 1961-95

(millions of tons)

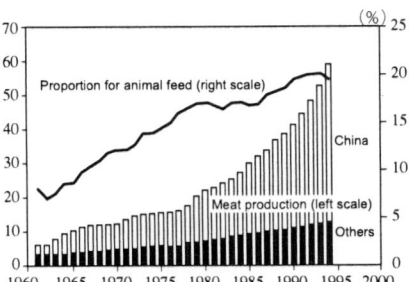

Source: FAO, *FAOSTAT Agricultural Data*, 1996.

---

[1] See Lester Brown's books: *Tough Choices: Facing the challenge of Food Scarcity*, *Full House: Reassessing the Earth's Population Carrying Capacity*, and *Who Will Feed China: Wake-Up Call for a Small Planet.*

[2] Here East Asia means, in addition to the seven major countries (Taiwan is not included because of its absence from FAO statistics), Myanmar, Cambodia, Laos, Vietnam, Singapore, Mongolia, North Korea, and Hong Kong for a total of 15.

[3] FAO, *World Agriculture: Towards 2010*, 1995.

[4] Ibid.

agricultural factors such as soil degradation, water pollution, and contamination by agrochemicals, which act to restrict the further expansion and intensification of agriculture. Even if East Asian countries have the economic leeway to further increase food imports, whether world food production can stabilize and realize sustained growth is another matter.

Second, economic growth is pushing us farther up the food chain. That is to say, increased demand for livestock products means that a considerable portion of grain production becomes animal feed (Fig. 3). East Asian meat production soared from 5.3 million tons in 1961 to 59.6 million tons in 1994, with the increase being especially great in China. In conjunction with that, the proportion of grain demand used for animal feed rose from 8% to 20%. This increase in grain demand for feed is behind the Worldwatch Institute's prediction that China's 2030 grain imports will reach 207 to 369 million tons, and the prediction of the U.S. Department of Agriculture, which, while critical of the Worldwatch prediction as an overestimate, says that China's

net grain imports in 2005 will be 32 million tons.[5]

On this basis Lester Brown observes that there will be a food catastrophe, and ways to avoid this come down to stabilizing the population. Brown contends that assuring food security requires that we push political leaders of countries and international organizations toward making new decisions, but his arguments lack critical analyses of the historical and structural factors that brought about this situation in the first place: a world trade system that consolidates the North-South differential, and the world food strategies of the U.S. government and agribusiness transnationals, which shackle the independent initiatives of countries and regions to increase their food self-sufficiency.

Just as with environmental and resource problems, agricultural and food problems must be discussed in a framework that is not technological, but rather political and economic.

(HISANO Shuji)

5 F.W. Crook and W.H. Colby, "The Future of China's Grain Market," *USDA Agricultural Information Bulletin*, No. 730, October 1996.

Table 1 Asia's Agriculture and Food Production

		1 China	2 Indonesia	3 Japan	4 S. Korea	5 Malaysia	6 Philippines	7 Thailand	8 Taiwan
A. Percentage of population engaged	1990	67.5	48.5	6.4	24.6	32.1	46.8	64.3	–
in farming, 1990 and future	2000	59.8	39.8	3.6	16.0	23.8	41.8	57.1	–
	2010	51.7	31.7	2.0	10.1	17.1	36.9	49.7	–
B. Cropland (1,000 ha)	1994	95,782	30,171	4,422	2,055	7,604	9,190	20,800	–
C. Grain-producing area (1,000 ha)	1994	87,904	13,843	2,448	1,220	733	6,657	10,515	–
D. Irrigated area (ha)	1970	38,123	3,900	3,397	1,187	257	830	1,965	–
	1980	45,301	4,209	3,056	1,309	323	1,218	3,007	–
	1990	47,232	4,410	2,846	1,344	337	1,560	4,248	–
E. Rice yield per unit area (kg/ha)	1970	3,281	2,346	5,485	4,628	2,397	1,683	1,933	–
	1980	4,236	3,262	5,587	5,512	2,844	2,205	1,888	–
	1990	5,625	4,298	6,120	6,231	2,842	2,778	2,034	–
F. Per capita grain production (kg)	1970	207	145	134	216	123	167	356	–
	1980	255	179	98	183	117	191	355	–
	1990	301	226	90	157	88	201	349	–
G. Net grain trade (exports-imports,	1970	-3,515	-1,069	-14,294	-2,628	-980	-816	3,105	–
1,000 t)	1980	-15,699	-3,067	-24,375	-5,902	-1,578	-891	5,695	–
	1990	-14,698	-2,156	-28,185	-9,874	-2,862	-2,204	6,705	–
H. Grain self-sufficiency rate (%)	1970	97.6	94.4	48.5	76.7	57.9	93.8	154.5	–
	1980	94.1	92.9	31.0	55.4	50.8	91.2	151.9	–
	1990	97.2	95.0	28.0	41.2	36.6	86.9	137.8	–
I. Meat production (1,000 t)	1970	7,811	414	1,616	167	160	568	544	–
	1980	14,239	619	3,002	471	289	785	845	–
	1990	30,309	1,353	3,499	924	626	1,098	1,255	–
J. Livestock feed proportion of domestic	1970	10.6	1.9	35.6	6.2	11.3	20.1	5.4	–
grain demand (%)	1980	16.0	2.1	46.0	17.5	22.1	20.8	11.4	–
	1990	18.3	5.8	46.8	33.5	36.7	22.7	26.2	–
K. Per capita daily food energy supply	1970	2,004	1,870	2,687	2,820	2,528	1,765	2,182	–
(cal)	1980	2,333	2,201	2,748	3,121	2,725	2,220	2,218	–
	1990	2,657	2,561	2,899	3,219	2,710	2,350	2,247	–

Notes: B: In addition to cropland, land under cultivation ordinarily includes land planted with permanent crops, temporary meadows, market and kitchen gardens, and temporarily fallow land. Permanent pastureland is excluded from agricultural land. D-K: Values are averaged over three years with listed year in the middle. H: Grain self-sufficiency rate = grain production / domestic demand for all uses. J: Grain uses include food, intermediate use (animal feed and seeds), non-food industrial uses, and natural decrease.

Sources: A: FAO (N. Alexandratos, ed.), *World Agriculture: Towards 2010* (FAO, Rome, 1995). C, E-F, I: FAO, *FAOSTAT Agricultural Data: Production* (FAO, electronic database, updated February 1997). B, D: FAO, *FAOSTAT Agricultural Data: Land* (FAO, electronic database, updated October 1996). G-H, J-K: FAO, *FAOSTAT Agricultural Data: Commodity Balances* (FAO, electronic database, updated July 1996).

[6] Pesticides, Chemical Fertilizers, and Eco-Agriculture

There is no denying the role of the green revolution, which in the postwar years and especially since the 1960s has achieved a measure of improvement in Asia's food supply and demand, but it is also a fact that despite the puffery about its achievements, the green revolution is not an unmitigated good. Environmental problems include soil salinization by irrigation, loss of species diversity due to dissemination of high-yield varieties, and soil and water contamination by the heavy application of chemical fertilizers and pesticides, and there are also social problems such as farm financing difficulties caused by rising production costs, and unequal inter-regional and inter-class income distribution.

Although chemical fertilizer use differs according to country, the increase between 1961 and 1994 per hectare of farmland was 14 kg to 618 kg in China, 5 to 85 in Indonesia, 19 to 159 in Malaysia, and 2 to 62 in Thailand, for enormous increases that contrast with Japan's slight decline (Fig. 1).

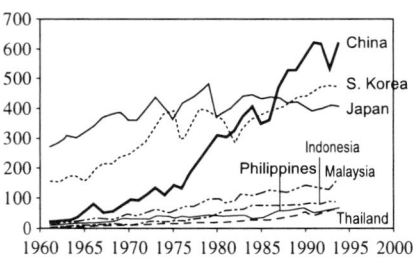

Fig.1 Chemical Fertilizer Inputs on Farmland in Seven Major East Asian Countries, 1961-95

Source: FAO, *FAOSTAT Agricultural Data*, 1996.

Unlike pesticides, the manufacture of chemical fertilizers requires no sophisticated technologies, allowing even developing countries to manufacture them domestically, but except for South Korea and Indonesia, which export more chemical fertilizers than they import, developing countries' imports are increasing. Further, their burdens for purchasing these fertilizers are growing, with 1994 imports by China amounting to $2 billion, and those by Thailand coming to $500 million.

In the 1960s and 1970s the use of pesticides also soared, and even now the Asian market continues to grow despite increasingly stringent worldwide controls. The East Asian market in 1974 was only 9.3% of the world market, but by the mid-1980s it had exceeded 20%, and in 1994 it was 26.6%

(Fig. 2). While Japan accounts for more than half of pesticide exports to other East Asian countries, Western agricultural companies have accorded East Asia a strategic place as a region with promise for greater future demand.

Meanwhile, there are international efforts to restrict the use, import, and export of agricultural chemicals. In 1985

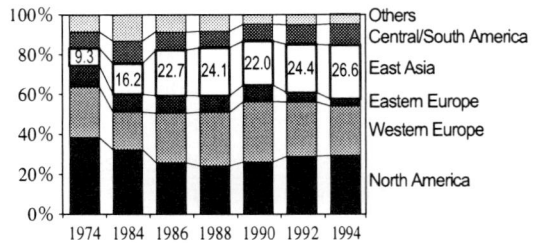

Fig.2 East Asian Share of Sales in World Pesticide Market

Source: Multiple issues of *Kagaku Keizai* (based on data from Wood Mackenzie Co.)

FAO adopted an international code of conduct on the distribution and use of pesticides, and in 1989 that code incorporated a prior informed consent system that made it possible to prohibit pesticide imports even if a country has no such domestic provisions. Nevertheless only 18 hazardous chemical substances (12 of them pesticides) were prohibited after 1991 when this system went into effect. As this shows, problems of this system include: Only a few of the pesticides used worldwide are subject to prohibition, and because the system is not legally binding, it is not possible to adequately restrict pesticides not subject to domestic laws.

In the West there are strict controls on pesticides at every stage from development to final use, and companies engaging in manufacture and distribution are expected to discharge their responsibilities to society. By contrast, there has been no mitigation whatsoever in the frequent pesticide-related accidents by agricultural producers in developing countries. These countries need implementation of Agenda 21's Chapter 19,[1] as well as emphasis on progress in talks by UNEP, ILO, WHO, and other involved organizations.[2]

There are also attempts at practicing eco-agriculture that does not depend on pesticides or chemical fertilizers. One of these is integrated pest management (Table 1), which in-

---

[1] "Environmentally Sound Management of Toxic Chemicals, Including Prevention of Illegal International Traffic in Toxic and Dangerous Products."

Table 1 *Effects of IPM Programs on Pesticide Use (Rice)*

Country	Change in agricultural use	Change in yield	Annual savings
Thailand	50%	n.d.	$5-10M
Philippines	62%	110%	$5-10M
Indonesia	34-42%	105%	$50-100M
China	46-80%	110%	$400,000
Vietnam	57%	107%	$54,000
Bangladesh	0-25%	113-124%	n.d.
India	33%	108%	$790,000
Sri Lanka	26%	135%	$1M

Source: Jules N. Pretty, *Regenerating Agriculture: Politics and Practice for Sustainability and Self-Reliance*, Joseph Henry Press, 1995, p. 98 (based on P. Kenmore, *How Rice Farmers Clean up the Environment, Conserve Biodiversity, Raise More Food, Make Higher Profits: Indonesia's IPM-A Model for Asia*, FAO, 1991, and other sources).

volves controlling pests with the integrated use of diverse methods including creative fertilization and tillage, crop rotation, mixed cultivation, thorough advance scouting of fields, and use of resistant crop varieties and biological pest

[2] Although it was international aid organizations and developing country governments that promoted the green revolution and used subsidies to lead farmers into pesticide-dependent agriculture, pesticides are certainly not a given. It is transnational corporations that develop, manufacture, and sell pesticides in the developing world market. Instead of allowing international regulation to be nothing more than a mere political slogan, we must push for the development of a code of conduct for transnationals, and the reinforcement of public controls in accordance with that code. Controls that are domestically possible in the West should be possible internationally as well.

controls. IPM is attracting interest as a farming method because it is sustainable environmentally, and also because it reduces the cost of agricultural supplies. Well known is the success of Indonesia, which began implementing IPM as a national project in 1986, substantially reducing the use of pesticides, and also saving money by cutting subsidies for those chemicals.[3]

Another attempt at eco-agriculture is the grassroots practice of organic agriculture. In 1993 the International Federation of Organic Agriculture Movements held its Asian conference in Japan. IFOAM works to reassess each region's traditional farming methods as eco-agricultural practices that can lead away from the chemical dependence established by the green revolution.

Both of the above initiatives place emphasis on weaning farmers off dependence on commercialized technologies, and on using education to put both producers and consumers in the driver's seat. Of crucial importance now is increasing the public research budget for agriculture, which is smaller than the amounts of money going to pay for farming supply imports and subsidies, and redirecting the money toward eco-agriculture research and practice.

(HISANO Shuji)

[3] Some pesticide companies are attempting to subsume IPM programs into their corporate strategies, making it necessary to monitor these attempts to see if they actually lead to the promotion of eco-agriculture.

Table 2 Pesticide and Fertilizer Use in Asia

		1 China	2 Indonesia	3 Japan	4 S. Korea	5 Malaysia	6 Philippines	7 Thailand	8 Taiwan
A. Land under cultivation (1,000 ha)	1994	95,782	30,171	4,422	2,055	7,604	9,190	20,800	–
B. Chemical fertilizer production	1970	5,228	44	2,744	532	33	94	10	–
(1,000 t)	1980	24,284	1,112	1,962	1,149	31	75	0	–
	1990	38,455	2,887	1,378	971	262	324	0	–
C. Chemical fertilizer use (1,000 t)	1970	8,298	215	1,993	565	172	204	105	–
	1980	29,046	1,161	2,013	808	429	332	293	–
	1990	54,803	2,380	1,847	929	916	521	897	–
D. Chemical fertilizer use per unit	1970	81	8	365	247	39	29	8	–
tilled land (kg/ha)	1980	289	45	412	368	89	38	16	–
	1990	568	78	402	441	133	57	44	–
E. Chemical fertilizer exports and	1970	-202.0	-28.3	-3.8	-1.0	-13.1	-9.6	-14.1	–
imports (exports minus imports,	1980	-784.4	-105.0	69.9	213.5	-156.2	-116.3	-148.3	–
millions of dollars)	1990	-2,763.4	127.7	-297.5	72.0	-132.6	-49.8	-409.7	–
F. Pesticide use (t)	1990-93	240,000	2,100	–	26,400	40,000	–	36,000	–
G. Pesticide exports and imports	1970	-25.0	-10.9	4.5	-1.1	-5.4	-3.9	-8.4	–
(exports minus imports, millions	1980	-61.2	-26.2	81.5	-10.7	-32.9	-13.2	-64.0	–
of dollars)	1990	-206.2	7.6	87.8	-9.3	-11.9	-28.6	-107.7	–
H. Number of tractors in use (thousands)	1970	153	9	278	0	4	7	6	–
	1980	738	9	1,327	3	8	11	18	–
	1990	829	29	2,052	42	20	10	58	–
I. Agricultural machinery exports and	1970	-28.0	-19.3	38.4	-2.6	-6.3	-14.9	-20.6	–
imports (exports minus imports,	1980	-61.5	-37.5	825.8	-29.6	-35.2	-29.4	-103.5	–
millions of dollars)	1990	-78.8	-37.9	575.1	-197.6	-37.6	-11.1	-198.5	–
J. Public agricultural research	1961-65	485.4	73.9	780.6	25.7	71.2	41.9	55.5	33.8
expenditures (millions of dollars)	1971-75	689.3	114.9	891.2	28.0	74.1	35.4	56.3	57.1
	1981-85	933.7	141.1	1,021.6	50.0	110.8	28.6	77.8	72.0
K. Number of full-time researchers	1961-65	11,563	450	13,798	887	295	973	585	770
	1971-75	20,048	1,005	13,747	1,052	663	1,390	1,343	1,142
	1981-85	32,224	1,349	14,779	1,356	811	1,965	1,676	1,607

Notes: A: See the notes to Table 1, "5. Trends in Agriculture and Food Production" for a definition of land under cultivation. B-E, G-I: Values are averaged over three years with listed year in the middle. F: Amounts of active ingredients, average from 1990 to 1993. J: Research expenditures by public, semi-governmental, scientific agricultural research institutes, etc. Data were collected in each country's currency, deflated with 1980 as the base year, and converted to dollars based on 1980 purchasing parity. K: Researchers with qualifications equal to bachelor degrees or higher who engage in agricultural research at public, semi-governmental, and scientific research institutes, and the like. Researchers from abroad also included. J-K: Averages for those periods of time.

Sources: A: FAO, *FAOSTAT Agricultural Data: Land* (FAO, electronic database, updated October 1996); B-I: *FAOSTAT Agricultural Data: Means of Production* (FAO, electronic database, updated August 1996); J-K: Pardey, P.G., J. Roseboom and J.R. Anderson, eds., *Agricultural Research Policy, International Quantitative Perspectives*, Cambridge University Press, 1991.

[7] Deforestation and Forest Conservation

Forests are defined by the FAO as ecosystems where the crowns of trees or bamboo cover at least 10% (in developing countries) or at least 20% (in developed countries) of the land, generally have wild animals and plants, and are not used for agriculture.[1] Currently the world has 3,443,000,000 ha of forested land, which is 27% of the world's land area. Tropical forests account for 1,756,000,000 ha, or over half of all the world's forests.

Accounting for most terrestrial plant biomass, forests are extremely valuable as a huge repository of not only carbon, but also as many as 10 million species. Tropical forests alone make up nearly half of all forest biomass, and they are home to between 2 and 5 million species.[2] Forests are therefore valuable for maintaining the global environment, for understanding the evolution of life on Earth by means of the genes they harbor, and for keeping as yet unknown resources that we can use.

But now the forests, which are essential to our survival, are disappearing fast. Between 1981 and 1990 the world's forests decreased by 16,270,000 ha, of which 15,410,000 ha were tropical. The breakdown is 4.1 million ha in Africa, 3.9 million ha in the Asia-Pacific, and 7.41 million ha in Latin America and the Caribbean countries.[3] Although Asia's loss does not equal that of Latin America and Africa, Asian tropical forests have the highest biodiversity, raising concerns about the loss of valuable genetic resources. During the last decade Indonesia lost a sizable area of its forests, while Thailand and the Philippines have but little remaining. China has very little forested land, and a desertification problem.

Deforestation is believed to be caused by non-traditional swidden agriculture, by people who make their way into the forests along logging roads left after commercial or illegal logging, and by conversion of forest land, especially the development of farmland and pastureland. Compared with Indonesia and Malaysia, which have high proportions of forested land, the high proportions of tilled land in China, the Philippines, and Thailand make for low proportions of forested land (Fig. 1).

Maintaining forests with value as genetic resources requires both sustainable development and efforts directed at suitable conservation. Some specific actions might be (1)

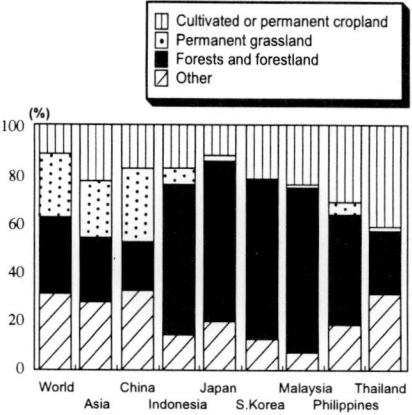

Fig.1 Types of Land Use

Source: FAO, *FAO Yearbook Production*, vol. 49, 1995.

managing some forests and logged areas as production forests, (2) permanent protection for as yet unlogged forests, and (3) regenerating forests in deforested areas in order that the original ecosystems may recover.

Let us consider specific actions that each of these goals would require.

(1) Governments must clearly and legally demarcate areas according to land use.

(2) One means would be to establish protected areas including national parks and the like. In 1990 there were 6,940 protected areas covering 651,460,000 ha worldwide. Since the 1970s protected areas have increased quickly in terms of both numbers and size. In Asia the numbers and total area of protected regions are large, especially in China and Indonesia, but even if protected marine areas are in-

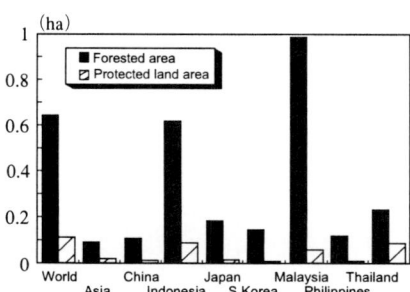

Fig.2 Per Capita Forested and Protected Land

Sources: FAO, *Forest Resources Assessment Global Synthesis*, 1990; World Conservation Union (IUCN), *United Nations List of National Parks and Protected Areas*, 1990.

---

[1]   FAO, *Forest Resources Assessment 1990 Global Synthesis*, 1995.
[2]   FAO, *Forest Resources Assessment 1990 Tropical Countries*, 1993.
[3]   May, Robert M. "How Many Species Inhabit the Earth?" *Scientific American*, vol. 267, no. 4, pp. 18-24, 1992.

cluded, the per capita size of protected areas is still far below that of forested land (Fig. 2). It will be necessary to

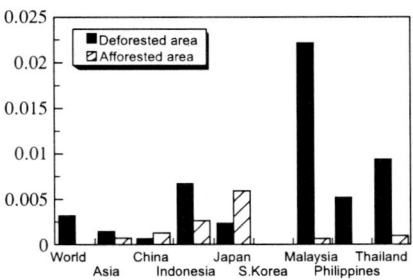

Fig.3 Per Capita Average Annual Areas Deforested and Afforested

Sources: FAO, *Forest Resources Assessment Global Synthesis*, 1990; FAO, *Forest Resources Assessment Tropical Countries*, 1990 (there were no data on the worldwide afforested area).

conserve biodiversity by continuing to increase protected areas and by managing them properly. Further, if countries adequately consider the livelihood rights of the indigenous peoples who have lived in protected areas and coexisted with the forests, it will be possible to adopt comprehensive protection and management policies.

(3) For this purpose a variety of afforestation projects is possible. In Asia there are both afforestation projects by Asian governments, and through aid provided by other governments, although at present they cannot keep up with deforestation (Fig. 3). In recent years there are attempts at participatory social forestry that involves local people. Asian countries will need to plant tree species adapted to each region in order to recover ecosystems, and carry out afforestation projects that accommodate the needs of local inhabitants.

(HARADA Kazuhiro)

Table 1 Forest Loss and Conservation (1990)

	1	2	3	4	5	6	7
	China	Indonesia	Japan	S. Korea	Malaysia	Philippines	Thailand
A. Total area (thousands of ha)	959,696	190,457	37,780	9,926	32,975	30,000	51,312
B. Land area (thousands of ha)	932,641	181,157	37,652	9,873	32,855	29,817	51,089
C. Forest and forestland area (thousands of ha)	162,029	145,108	25,146	6,459	22,248	13,640	14,968
Forest area (thousands of ha)	133,799	115,674	23,780	6,281	17,664	8,034	13,264
Natural forest (thousands of ha)	101,968	109,549	13,382	6,281	17,583	7,831	12,735
Plantations (thousands of ha)	31,831	6,125	10,398	0	81	203	529
Proportion of forested land (%)	14	64	63	64	54	27	26
Forestland area (thousands of ha)	28,230	29,434	1,366	178	4,584	5,606	1,704
D. Average annual natural forest loss, 1981-1990 (thousands of ha)	400	1,212	284	-1	396	316	515
Natural forest loss rate, 1981-1990 (%)	–	1.0	0.02	–	2.0	3.3	3.3
E. Average annual afforestation increase, 1981-1990 (thousands of ha)	1,139.8	474.0	706.5	0	9.0	-1.0	42.0
Afforestation rate, 1981-1990 (%)	–	8.1	7.2	n.s	16.1	-0.3	8.5
F. Forest biomass (tons/ha)	157	203	62	120	261	236	125
Forest biomass (millions of tons)	16,009.6	22,261.4	1,498.0	754.9	4,590.9	1,848.4	1,585.3
G. Number of protected zones	289	169	65	17	45	28	83
Area of protected zones (thousands of ha)	21,947	17,799	2,402	577	1,162	583	5,105

Notes: C3: 1995 data; D3: 1986-1995 data; E3: 1985-1994 data; proportion of forested land is forested area divided by land area; n.s.: not significant.

Sources: A-F: FAO, *Forest Resources Assessment 1990 Global Synthesis* (1995); D and E data for Indonesia, Malaysia, Philippines, and Thailand are from FAO, *Forest Resources Assessment 1990 Tropical Countries* (1993); C3-E3: Forestry Agency, ed., *Forestry Statistics Handbook 1988* (1996) ; G: IUCN, *United Nations List of National Parks and Protected Areas*, 1990.

[8] Regionally Disproportionate Wood Production and Consumption

Seen worldwide, the loss of forests has a variety of causes. A major cause is a lack of regional balance in wood production and consumption.

Log production has increased over 50% from 2.2 billion m³ in 1965 to 3.4 billion m³ in 1993 (Table 1). Of this, 1.5 billion m³ of logs go to industrial use, much of which becomes lumber and plywood.

The U.S. is the largest producer of industrial logs, followed by Canada and Russia (together accounting for 170 million m³). Large Asian producers are China, Malaysia, Indonesia, and some other countries with vast land areas. Although the Philippines used to be a big producer, due to depleted forest resources its production has declined rapidly from 15 million m³ in 1970 to 9 million m³ in 1980, and further to 4 million m³ in 1993.

Softwood and hardwood logs have different characteristics and therefore different uses. Production of softwood industrial logs, which are mostly processed into lumber, was 930 million m³, of which 42 million m³, or approximately 4%, went to international trade in 1993. Big exporters are the U.S., Russia, and some other countries, while Japan and South Korea import the most.

As old-growth forests dwindled in the U.S., which has exported many softwood logs, the latter half of the 1980s saw the rise of environmental movements symbolized by issues such as protection of the spotted owl. This led to logging restrictions starting in the late 1980s, and with added domestic demands for protection of the forest products industry, controls on log exports were tightened in 1990.

Of 530 million m³ of hardwood industrial logs produced, 3% or 17 million m³ were traded worldwide in 1993. Currently Malaysia has the highest log exports, with most of them imported by Japan and China.

Although most hardwood logs are tropical, the main producing countries changed, in accordance with importers' quest for high-quality wood resources, beginning with the Philippines in the 1960s to the mid-1970s, then shifting to Indonesia from the 1970s to the 1980s, and then to Malaysia from the 1980s.

Meanwhile, the full emergence of "resource nationalism" in Southeast Asian countries brought Indonesia to prohibit log exports in 1985 in order to promote domestic plywood production and export. In the 1990s Malaysia's Sabah and Sarawak states also limited log exports.

Worldwide lumber production in 1993 was 430 million m³, of which 310 million m³ were softwood and 120 million m³ were hardwood. Much softwood lumber production was by the U.S. and Canada, which have plentiful softwood resources, while most hardwood lumber was produced by the U.S. and India.

Since the 1970s Canada has promoted lumber exports and is the largest exporter of such products, most of which go to the U.S. and Japan.

Trade in hardwood lumber was 17 million m³ in 1993, and the largest importer is Thailand, which began losing its forests early. The biggest exporters are Malaysia and the U.S., which have comparatively extensive hardwood resources.

Production of plywood, which is used mainly in concrete forms and other building applications, was 48 million m³ in 1993, with the U.S. and Indonesia being the big producers. Indonesia and Malaysia export much plywood, most of which is imported by Japan and China. The chief producer of plywood was Japan until the mid-1970s, then South

Fig.1 Japan's Log Imports

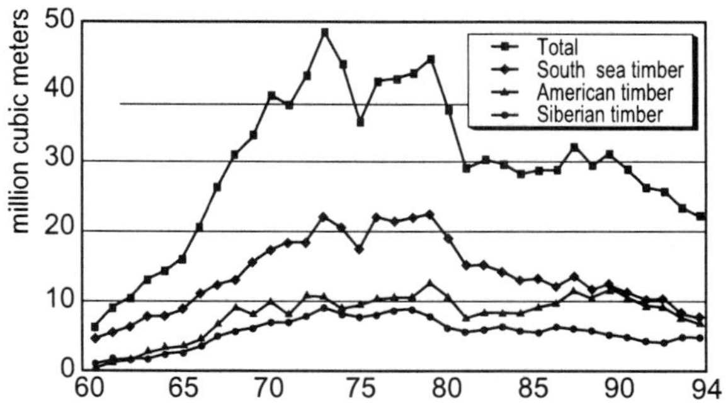

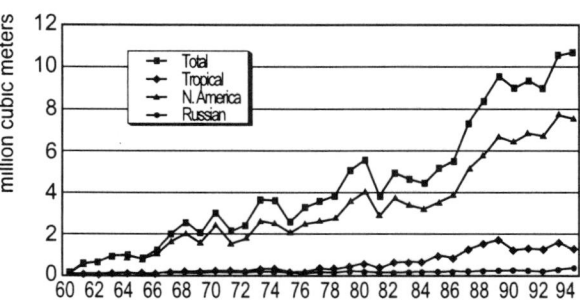

Source: Ministry of Finance, *Trade Statistics*, various years.

Fig.2 Japanese Imports of Lumber

Source: same as Fig.1.

Korea and Taiwan, and since the mid-1980s it has been Indonesia owing to its log export ban.

Paper consumption is said to be an indicator of a country's economic level. Production of paper and paperboard was 250 million tons in 1993, mainly in the consuming countries of the U.S. and Japan, and one-fourth of that was traded. Northern European countries are the major exporters, and the U.S. imports a great deal.

Recent trends in the wood trade show that, owing to the increasing seriousness of environmental problems and the rise of conservation movements, as well as the dwindling of forest resources, there is a clear shift from trade in logs to that in wood products.

That change is reflected well in Japan's wood trade. Despite Japan's abundant forest resources, it imports large amounts of wood. With logs, for example, imports of so-called "south sea timber" from the Philippines, Indonesia, Malaysia, and other Southeast Asian countries peaked in the 1970s, and imports of North American logs from the U.S. and Canada peaked in the late 1980s, subsequently declining (Fig. 1). On the other hand, imports of lumber and processed items are increasing. Those of North American products, mainly lumber, have tended to increase since the 1960s, and since the second half of the 1980s have grown rapidly. Imports of south sea timber products, mainly plywood, also shot up starting in the 1980s (Fig. 2).

Wood is a major item of world trade, and making sustainable use of forest resources will necessitate a system that takes a new view of production and consumption from the perspective of the wood trade, and balances use with the growth of forests.

(TACHIBANA Satoshi)

Table 1 World Wood Production and Trade

	World	Asian eight-country total	1 China	2 Indonesia	3 Japan	4 S. Korea	5 Malaysia	6 Philippines	7 Thailand	8 Taiwan	9 U.S.A.	10 Canada
A. Roundwood production (thousands of m³)	3,404,413	659,788	300,668	188,118	32,570	6,485	54,332	39,576	38,039	–	495,800(F)	179,967(F)
Roundwood imports	111,416	62,382	6,184	60	45,489	8,672	72	539	1,366	–	2,460	5,030
Roundwood exports	109,969	14,014	2,089	1,632	57	9	9,626	349	252	–	26,680	3,961
Coniferous roundwood production	1,128,567	167,984	144,258(F)	539(F)	18,772	4,313(F)	100(F)	2(F)	–	–	303,000(F)	167,032
Non-coniferous roundwood production	2059,130	475,227	154,470(F)	186,619(F)	6,949	2,070(F)	51,652(F)	39,214	34,253	–	189,800(F)	12,935(F)
B. Fuelwood and Charcoal production	1,875,855	434,643	200,060(F)	149,063(F)	361	4,491(F)	9,375(F)	35,980(F)	35,313(F)	–	93,300(F)	6,834(F)
C. Industrial roundwood production	1,528,557	225,144	100,608	39,054(F)	32,209	1,994(F)	44,957(F)	3,596	2,726	–	402,500(F)	173,133
Industrial roundwood imports	106,310	61,395	5,920	273	44,895	8,528	17	538	1,224	–	1,809	4,957
Industrial roundwood exports	104,633	12,360	1,927	704	47	6	9,517	20	139	–	26,419	3,576
Share of coniferous industrial round-wood production	929,254	84,110	63,281(F)	539(F)	18,772	1,416(F)	100(F)	2(F)	–	–	285,800(F)	165,275
Coniferous industrial roundwood imports	41,497	20,236	–	2	14,780	5,409	–	43	2	–	456	3,359
Coniferous industrial roundwood exports	42,320	273	178	1	7	–	86	1	–	–	15,381	1,821
Share of non-coniferous industrial round-wood production	528,915	132,516	35,447(F)	38,516(F)	6,798	578(F)	44,857(F)	3,594	2,726	–	116,700(F)	7,858(F)
Non-coniferous industrial round-wood imports	17,328	13,853	2,596	7	7,531	2,088	–	417	1,214	–	7	1
Non-coniferous industrial round-wood exports	16,838	9,444	–	42	–	–	9,382	2	18	–	–	–
D. Sawlogs and veneer logs production	899,429	150,587	51,769(F)	35,833(F)	17,567	1,066(F)	43,600(F)	687	65	–	241,700(F)	137,159(F)
Share of coniferous sawlogs and veneer logs production	605,091	49,945	32,733(F)	333(F)	16,099	678(F)	100(F)	2(F)	–	–	185,500(F)	132,566(F)
Share of non-coniferous sawlogs and veneer logs production	294,338	100,642	19,036(F)	35,500(F)	1,468	388(F)	43,500(F)	685	65	–	56,200(F)	4,593(F)
E. Sawnwood and sleepers production	432,410	73,655	25,268(F)	8,338	26,260(F)	3,199	9,385(F)	490	715	–	106,167	59,774
Sawnwood and sleepers imports	98,788	16,595	2,521	14	10,626	1,198	39	290	1,907	–	36,489	1,460
Sawnwood and sleepers exports	102,202	6,956	653	672	27	72	5,362	103	67	–	9,411	43,555
F. Plywood production	48,343	21,940	2,638	10,050	5,263	898	2,500(F)	331(F)	260	–	17,094	1,824
Plywood imports	16,527	7,770	2,510	12	4,105	1,105	11	3	24	–	1,662	288
Plywood exports	17,072	11,668	247	8,904	30	22	2,421	39	5	–	1,567	416
G. Paper and paperboard production (thousand metric tons)	253,586	62,440	23,816(F)	2,600(F)	27,764	5,804	663	487	1,306	–	77,250	17,557
Paper and paperboard imports	63,099	6,710	3,056	177	1,482	505	617	296	577	–	11,885	1,246
Paper and paperboard exports	64,483	3,810	984	669	1,288	699	47	2	121	–	7,146	12,896

Note: Figures followed by (F) are FAO estimates.
Source: FAO, FAO Yearbook Forest Products 1993.

[9] Rapidly Increasing Marine Catches

The world fish catch in 1989 (including inland waters and aquaculture) reached an all-time high of 100,120,000 tons, remained steady until 1992, and then set a new record of 101,420,000 tons in 1993. But there is little doubt that the catch is fast approaching nature's production limit (Fig. 1).

Eight Asian countries — China, Indonesia, Japan, South Korea, Malaysia, the Philippines, Thailand, and Taiwan — together have rapidly increasing catches (Fig. 2). Their share of the total world catch rose from 27% in 1970 to 39% in 1993, and the latter year's figures indicate that the six countries excluding Malaysia and Taiwan were among the world's 12 countries with the highest catches. China's catch in particular has skyrocketed, jumping 5.7-fold from 3.1 million tons in 1970 to 17.5 million tons in 1993. China eclipsed Japan in 1989 to become the world's biggest fisher, while Japan's catch has since 1989 steadily declined owing to factors such as the slide in the sardine stock.

Reasons for the rising catch of these eight Asian countries are increases in their shore and adjacent sea fisheries, and in pelagic fishing by South Korea and Taiwan, as well as expanding aquaculture production. From 1984 to 1993 the eight countries' aquaculture production surged 2.7-fold and increased aquaculture's share of their total marine catch (Fig. 3). Their aquaculture production accounts for 73% of the world total, which becomes 82% when India is added (1993).

The increase in marine catches seems desirable from the perspective of food production, but actually entails a number of problems.

First, over the long term the increase in fish taken hampers the sustainable use of fishery resources, therefore generating concern about future food production, and Asia's coastal fishery resources are suffering from overfishing and exhaustion.[1]

Second, aquaculture involves problems such as environmental damage caused by cutting mangrove forests and polluting water by building aquaculture ponds, adverse effects on biodiversity by introducing exotic species, competition with other sectors for land and water, and the waste of food resources through heavy fish feeding. Especially in

---

[1]   FAO, Inland Water Resources and Aquaculture Service, Fishery Resource Division, "Review of the State of World Fishery Resources: Inland Capture Fisheries," *FAO Fisheries Circular* 885, 1995.

Fig.1 World Fish Catch, 1970-1993

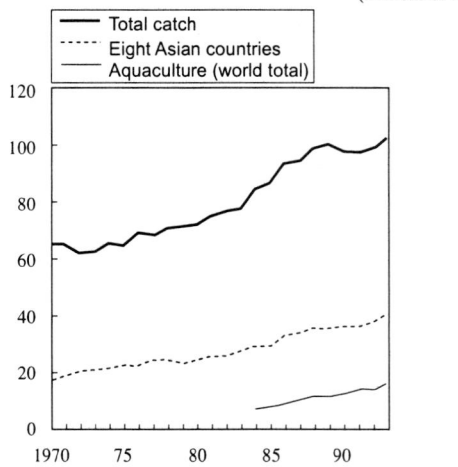

(millions of tons)

Source: FAO, etc. (*see* sources for Table 1.)

Fig.2 Catch Increases of Eight Asian Countries (growth rate when 1970=100)

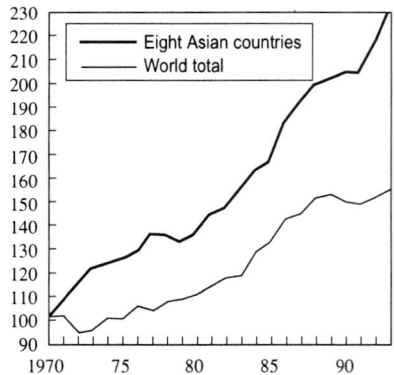

Source: Same as Fig.1.

Fig.3 Growth of Aquaculture Production and Fish Catches in Eight Asian Countries (1984=100)

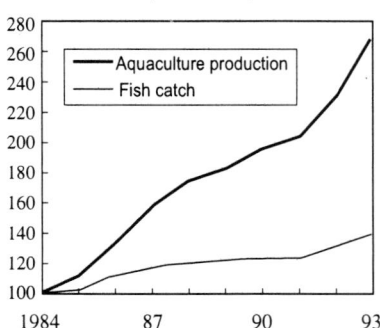

Source: Same as Fig.1.

the Asia-Pacific, the growth of aquaculture is said to be the largest cause of mangrove forest destruction.[2]

Third, the increase in fisheries catches does not necessarily translate into better nutrition. As discussed in Item 10, especially in countries like Indonesia and Thailand, marine products increasingly tend to become commodities on the international market instead of appearing on the people's dinner tables. In other countries as well, low-income citizens do not share adequately in the bounty of increased fishery production.

Fourth, a concomitant of the rising fish take is the huge bycatch. In many cases this is simply discarded, with the amount estimated at about 27 million tons annually. This discarded bycatch is largest in the shrimp trawling industry that operates chiefly in Southeast Asia and other tropical areas, and accounts for at least one-third of the total.[3]

It is often the case that problems such as those described above are caused not so much by Asia's fisheries industry overall as by a portion of the large-scale fisheries industry involved in the commercial trade in marine products, and

by intensive aquaculture.[4] Especially characteristic of sea fisheries in Southeast Asia is a dual arrangement in which a few large-scale commercial entities coexist with many small fishermen. The former run operations such as high-productivity shrimp trawlers, and take mainly mid- and high-priced species for export or urban markets, while the latter use low-productivity methods to take fish for either domestic provincial markets or for subsistence.[5] There is a face-off between the two groups over stocks, fisheries, income, and other issues. In aquaculture there is a differential between the intensive and the extensive and semi-extensive forms. During recent years there has been a reconsideration of development policy aimed primarily at large-scale fishery operations and intensive aquaculture, and attention is increasingly focused on community-based fisheries management.[6] Facilitating such a shift will require the revision of fishing laws to provide for fishing rights and fishing permit systems, as well as organizing fishing operations, but such efforts are still in their infancy.

(YOKEMOTO Masafumi)

[2] Economic and Social Commission for Asia and the Pacific (ESCAP) and Asian Development Bank (ADB), *State of the Environment in Asia and the Pacific 1995*, United Nations, New York, 1995, p. 126.

[3] Alverson, Dayton L., et al., "A Global Assessment of Fisheries Bycatch and Discards," *FAO Fisheries Technical Paper 339*, 1994.

[4] Greenpeace International, "Fishing and Food Security: A Greenpeace Perspective," presented to the International Conference on the Sustainable Contribution of Fisheries to Food Security, Kyoto, Japan, December 4-9, 1995.

[5] Smith, Ian R., "A Research Framework for Traditional Fisheries," *ICLARM Studies and Reviews* 2, 1979, pp. 3-4.

[6] FAO Fisheries Department, "Marine Fisheries and the Law of the Sea: A Decade of Change," *FAO Fisheries Circular 853*, 1993, pp. 44-46.

Table 1 Increases in Fish Catches and Aquaculture Production for Eight Asian Countries

	1 China	2 Indonesia	3 Japan	4 S. Korea	5 Malaysia	6 Philippines	7 Thailand	8 Taiwan
A. Fish catch (thousands of tons)								
1970	3,096.2	1,225.2	8,824.8	754.0	340.1	1,030.8	1,437.3	613.2
1980	4,235.3	1,841.8	10,434.0	2,091.1	736.1	1,555.7	1,798.0	936.3
1990	12,095.4	3,044.2	10,354.2	2,843.1	604.0	2,209.6	2,786.4	1,455.5
1993	17,567.9	3,637.7	8,128.1	2,649.0	680.0	2,263.8	3,348.1	1,424.0
B. World ranking of marine catch								
1970	4	12	2	17	31	16	8	21
1980	3	11	1	9	23	13	12	20
1990	1	8	3	9	25	11	10	18
1993	1	8	3	10	26	12	9	16
C. Aquaculture production (thousands of tons)								
1984	2,452.3	270.8	629.5	295.7	68.3	336.3	111.9	245.0
1993	8,880.2	592.1	833.0	391.4	105.2	391.7	414.3	285.3
D. Fish catches in inland waters (thousands of tons)								
1987	587.4	276.3	130.0	47.7	7.9	249.1	86.9	0.6
1992	1,232.8	301.8	96.7	25.0	3.3	234.8	122.0	0.8

Note: Aquaculture production (C) uses figures for "fish and shellfish."

Sources: A1-7: FAO, FAOSTAT-PC (v. 3.0) on diskette (FAO, Rome, 1995); B: FAO, *Yearbook of Fishery Statistics: Catches and Landings* (FAO, Rome, various years); C1-7: FAO Fishery Information, Data and Statistics Service, "Aquaculture Production Statistics 1984-1993, " *FAO Fisheries Circular 815* (Rev. 7), 1995; D: FAO, Inland Water Resources and Aquaculture Service, Fishery Resources Division, "Review of the State of the World Fishery Resources: Inland Capture Fisheries," *FAO Fisheries Circular 885*, 1995; A-C8: *Fisheries Yearbook Taiwan Area 1994* (Taiwan Fisheries Bureau, Department of Agriculture and Forestry, Provincial Government of Taiwan, 1995).

[10] Trends in Marine Product Consumption and Trade

Let us begin by using food balance sheets for fish and fishery products to examine trends in marine product consumption, imports, and exports during the 1980s in the seven Asian countries of China, Indonesia, Japan, South Korea, Malaysia, the Philippines, and Thailand (Table 1A). Here supply means the total marine catch plus imports, and demand means non-food uses, exports, and food supply (meaning the supply of marine products for human consumption), calculated as live weight fish.

Except for Japan and Malaysia, the supply of marine products for food and the supply per capita are increasing. Because the rate of increase for food supply in the seven countries as a whole is somewhat higher than in the world total, their share of the world supply grew slightly from 38% (1979-81 average) to 40% (1988-90 average). On the other hand, overall the ratio of food to non-food (animal feed, fertilizer, etc.) supply and/or exports in these seven countries is moving farther toward non-food uses. I shall therefore next examine pressure on the marine food supplies of these countries. Calculated in terms of live-weight catch and monetary value, the seven countries are divided into the following three groups.

(1) Deficits in both: Japan

(2) Catch deficits, value surpluses: Malaysia, the Philippines

(3) Surpluses in both: China, South Korea, Indonesia, Thailand

Japan has exceptionally large marine product imports both among the seven countries and also seen worldwide. In monetary terms Japan's imports account for 32% worldwide (1993), the world's highest. In the 1980s non-food demand increased its share of total demand 2.6-fold, while food supply dropped in terms of both absolute size and share.

In group two countries, Malaysia and the Philippines, deficits and surpluses changed places in catch and value, which was because, despite their differences in marine product categories and shares, they export shrimp and other relatively high-value marine products, while importing inexpensive products. Because much of these countries' imports appear to be for food, the main reason for pressure on their food supplies (on an unprocessed catch weight basis) is not exports but the supply for non-food purposes.

Group three countries can be further divided into two groups according to their import and export trends: A growing gap between imports and exports in both catch and value (Indonesia), and a narrowing gap (China, South Korea, and Thailand). Indonesia's imports are not increasing. In the 1980s the food supply component of its total demand dropped 5% while exports grew 4%, thereby pressuring the food supply. Meanwhile imports by China, South Korea, and Thailand are increasing more than exports. In Thailand, which is trying to achieve industrialization by fostering agro-industry, canned tuna exports are burgeoning, but this augments imports because Thailand cannot supply all the tuna it needs. Exports are mainly frozen shrimp and squid, and canned tuna, which together accounted for 63% of total growth in exports during the 1980s, and now account for 30% of total demand (1988-90 average). By contrast, the food supply is very low at 36%. Unlike Thailand, in China and South Korea the growth in imports is fueled by domestic consumption resulting from economic growth. Nevertheless, especially in South Korea the growing non-food supply is straining the food supply.

Finally let us explore the overall trend in the marine products trade. Exports of marine products from 1980 to 1993 by the seven countries and regions in U.S. dollars increased 3.3 times, while imports grew five-fold (Fig. 1). As a share of the total world trade, in value their exports grew from 18 to 22%, and imports from 21 to 37%. Thailand had the greatest increase in exports (9.5-fold) and China the greatest in imports (54.1-fold).

Changes since the 1980s in receiving countries and re-

Fig.1 Growth of Trade in Marine Products

(1980=100)

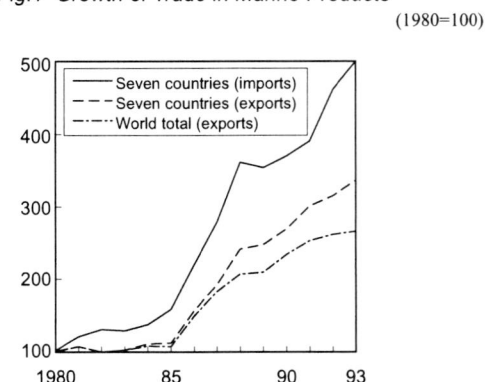

Source: FAO (*see* sources for Table 1.)

---

[1]   Indonesia, Malaysia, the Philippines, Thailand, Taiwan, Hong Kong, and Singapore

Fig.2 Change in Destinations for Nine Marine Products
Exported from the South China Sea Area

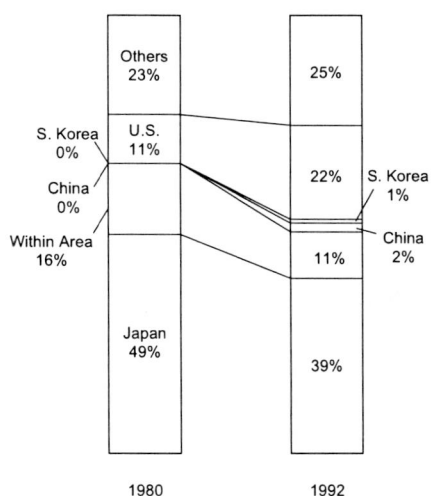

Source: SEAFDEC (*see* sources for Table 1.)

gions for nine marine product export items (in U.S. dollars) from the South China Sea region[1] show that new consumer nations have come on the scene (Fig. 2, Table 1D).

The reasons are that, first, the proportion of exports going to Japan has decreased. This is because there is expanding demand in the U.S. and other developed countries for shrimp (fresh, refrigerated, frozen), which are a considerable portion of exports.

Second, the value of exports to Japan of marine products other than shrimp, such as fish meal, is increasing, but as the share of shrimp in total export value is high, on the whole Japan's relative position as an importer is declining relative to other countries.

Third, consumption increases in China and South Korea are slightly augmenting the value of exports to those countries. In terms of absolute value, from 1980 to 1992 exports to China grew 112 times and those to South Korea 12 times.

(YOKEMOTO Masafumi)

Table 1 Supply, Demand, and Trade of Marine Products

	1 China	2 Indonesia	3 Japan	4 S. Korea	5 Malaysia	6 Philippines	7 Thailand	8 Taiwan
A. Food balance sheets of fish and fishery products (live weight, 1,000 t)								
1979-81 average								
Total supply	4,222.2	1,837.4	12,353.8	2,273.6	956.0	1,626.7	1,970.0	-
Production	4,222.2	1,830.2	10,347.6	2,206.5	745.6	1,572.8	1,909.4	-
Imports	-	7.2	2,006.2	67.1	210.4	53.9	60.6	-
Non-food uses	1.7	8.1	1,811.5	88.4	156.0	0.5	829.4	-
Exports	138.9	90.7	774.0	555.5	161.9	81.3	238.2	-
Food supply	4,081.6	1,738.6	9,768.4	1,629.7	638.1	1,544.9	902.4	-
1988-90 average								
Total supply	11,392.5	2,946.5	14,087.0	3,124.8	887.6	2,253.7	3,096.9	1,473.9
Production	11,224.7	2,939.3	11,164.4	2,798.1	607.4	2,106.0	2,671.8	1,386.5
Imports	167.8	7.2	2,922.6	326.7	280.2	147.7	425.1	87.4
Non-food uses	385.3	19.3	4,725.2	465.1	207.1	0	1,042.0	82.0
Exports	463.7	264.5	503.8	615.4	202.4	139.4	943.0	611.7
Food supply	10,543.5	2,664.1	8,857.9	2,044.3	479.0	2,114.3	1,117.4	781.6
B. Per capita supply of marine products for food (kg/yr)								
1979-81 average	4.1	11.6	83.0	41.7	44.3	30.6	18.8	-
1988-90 average	9.4	14.8	71.9	44.2	27.5	35.5	20.4	39.1
C. Marine product exports (1993, millions of US$)								
Exports	1,542.4	1,419.5	767.0	1,335.4	306.5	478.1	3,404.3	1,266.2
Imports	575.9	99.8	14,187.1	545.5	265.0	94.6	830.5	474.8
D. Value of exports of nine marine products								
1980								
Nine-item total (millions of US$)	-	187.6	-	-	102.7	69.1	332.6	-
Shrimps, prawns and lobster (fresh and frozen)	-	180.7	-	-	56.8	21.6	98.1	-
Ratio of exports to Japan	-	82%	-	-	34%	39%	36%	-
Ratio of exports to U.S.	-	2%	-	-	2%	39%	12%	-
1992								
Nine-item total (millions of US$)	-	1,255.7	-	-	189.2	356.4	3,035.7	371.2
Shrimps, prawns and lobster (fresh and frozen)	-	762.1	-	-	93.9	212.6	1,245.4	114.2
Ratio of exports to Japan	-	53%	-	-	26%	51%	35%	37%
Ratio of exports to U.S.	-	13%	-	-	13%	19%	28%	20%

Notes: A: 1) Excludes whale and seaweeds. 2) Total supply (production plus imports) is non-food uses plus exports plus food supply. However, both sides do not always coincide exactly. 3) Information on changes in stocks is available for a limited number of countries only. A-B8: Data for "Other Asia" items, considered about the same as those for Taiwan. C8: Data for 1992. Exclude the items "5. Miscellaneous products of aquatic animal origin" and "6. Products of aquatic plant origin." D: The nine items here are: 1) Fresh, chilled or frozen fish (excluding ornamental fish, fish fry, live fish, eels, and tuna), 2) dried, salted, or smoked fish (excluding shark's fin), 3) fresh and frozen shrimps, prawns and lobster, 4) fresh and frozen squid, cuttlefish and octopus, 5) dried and salted squid and octopus, 6) fresh, frozen, dried, and salted mollusks (excluding squid, cuttlefish, octopus, mussels, and arkshells , 7) fish products and preparations in airtight containers, 8) Crustacean and mollusc products and preparations in airtight containers, and 9) meals and similar animal feedingstuffs and fertilizers of aquatic animal origin.

Sources: A-B, C1-7: FAO, *Yearbook of Fishery Statistics: Commodities* (FAO, Rome, various years), C8, D: Southeast Asian Fisheries Development Center (SEAFDEC), *Fishery Statistical Bulletin for the South China Sea Area* (SEAFDEC Bangkok, various years).

[11] The Wildlife Trade in Asia

Human livelihood has long been dependent on wildlife, and even now we benefit from it. Trade in wildlife excluding timber and marine products has reached at least $10 billion annually.[1] Asia in particular not only makes heavy use of wildlife, but is also a supplier for the wildlife trade, making the sustainable use of wildlife a matter of the utmost importance to Asia.

CITES

The *Red List of Threatened Animals* issued in 1996 by the World Conservation Union (IUCN) lists a total of 5,205 species, of which 2,769 species, or about 53%, live in Asia. Factors driving them toward extinction include habitat deterioration, as well as overuse that also involves poaching. The 1973 Convention on International Trade in Endangered Species of Flora and Fauna was created to control this trade. Japan ratified it in 1980.

Japan — A Major Wildlife Consumer

Japan is one of the world's most active traders in wildlife, and a major consumer. Since 1985 Japan's imports of CITES-listed species have climbed rapidly, with 1994 imports being about 32 times those of 1981. In 1994 there

[1]   In 1994, according to TRAFFIC USA.

were far more imports than exports, the import-export ratio for that year being 97 to 3 (Fig. 1).

Nearly all the CITES species imported by Japan are listed on Appendix II. In 1994 these accounted for about 97.7% of all imports, and of that portion about 40.2% were reptiles and about 33% were plants. Approximately 99% of the reptile imports were manufactured products, and about 97% of the plants were living.

A look at the origins of CITES species imported in 1994 shows that Japan imported approximately 48% of the Appendix II species (including artificially bred and live items, as well as parts and derivatives) from other Asian countries. The biggest exporters to Japan are Hong Kong for mammals and Taiwan for birds, followed by Italy, China, and Portugal for reptiles. Manufactured products are the reason for high-volume imports from Italy. The biggest plant exporter to Japan is Thailand, followed by Taiwan.

While these are the countries exporting to Japan, they are not necessarily the exports' countries of origin. Determining where the CITES wildlife species imported by Japan ultimately come from requires that one investigate their countries of origin. In 1994 China was the largest source of mammals, and Taiwan of birds. And while all the birds imported from Taiwan are artificially bred, those from Indonesia are all wild. Sources of reptiles are the U.S., Columbia,

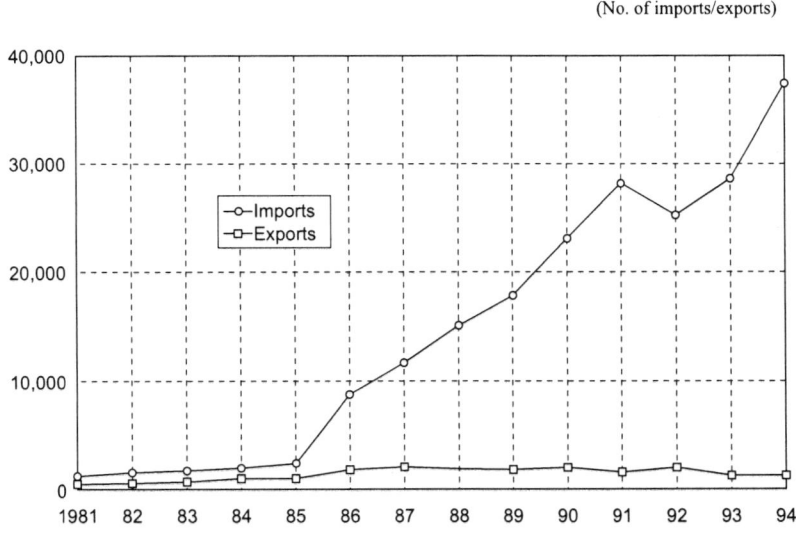

Fig. 1 *Imports and Exports of CITES Species by Japan*
(No. of imports/exports)

Source: Ministry of International Trade and Industry, *CITES Annual Report.*

Fig. 2 Sources of Japan's Appendix II Species Imports that Are of Wild Origin, 1994

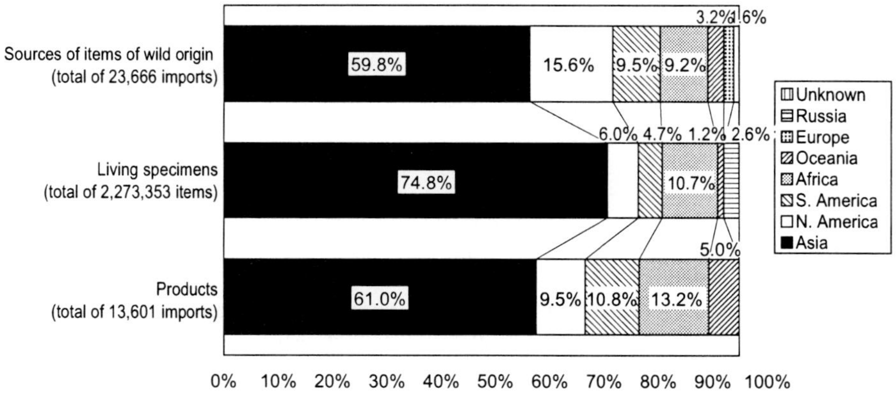

Source: Ministry of International Trade and Industry, *CITES Annual Report.*

and Zimbabwe, followed by fourth-place Indonesia. Nearly all are wild. In the case of plants the main countries of origin are also the main exporters, with Thailand first and Taiwan second. Fig. 2 shows imports to Japan of items specifically of wild origin.

Traditional Chinese Medicine

Characteristic of Asian trade in wildlife is the use of wild flora and fauna in traditional Chinese medicine. Behind this is the higher income level in parts of Asia, which allows increasing numbers of ordinary people to buy the expensive medicines once available only to the high-ranking and the well-heeled. The resulting increased demand puts pressure on rhinoceroses, musk deer, tigers, and the other endangered species that supply ingredients.

Especially reduced in number is the tiger, whose range states are Asian countries, and whose overuse is mostly in Asia. Tigers are classified into eight subspecies, and only a hundred years ago there were well over 100,000 of them throughout Asia. Since 1900, tiger populations have been reduced as much as 95%, and three subspecies have become extinct. The five surviving subspecies are greatly diminished, with the estimated number of individuals — 5,100 to 7,500 — putting the tiger on the threshold of extinction.

All parts of the tiger from nose to tail are used, but the bones in particular are highly valued in traditional medicines. Tiger bones are primarily used to strengthen human users' muscles and bones, and as painkillers for rheumatism and neuralgia. Although tigers were formerly poached mainly for their skins, the poachers' main object now is the bones, and this is one of the factors driving the tiger toward extinction.

A TRAFFIC study of 29 tiger range states and consumer nations indicates that only a few countries have the domestic laws considered necessary to protect tigers. With wildlife such as tigers, which cannot be protected with international trade restrictions alone, additional requirements are measures to conserve habitat and control poaching in range states, and domestic trade restrictions in consumer nations.

The current increase in human population and deterioration of habitat impose limits on the use of wildlife, which requires the formulation of measures to ensure that use is henceforth kept at sustainable levels.

(ISHIHARA Akiko)

[12] Biodiversity in Asia

Asia has some of the richest biodiversity in the world. East Asia, for example, has 14% of the world's mammal, bird, and fish species, about 26% of the ferns, and 40% of conifer species. Southeast Asia includes Indonesia, which can boast the world's largest number of mammal species; Malaysia, which has two of the world's tropical forest hotspots (the Malay Peninsula and Northern Borneo); and the marine region from the Indian Ocean to the West Pacific, said to have the world's most abundant coral reefs and mangrove forests (Fig. 1).

But Asia also surpasses other world regions in its spi-

raling population and its economic growth. China, Indonesia, and Malaysia are known for their biological megadiversity, but they also have population densities to match, and that is why Asia is fast losing its biodiversity.

During the decade of 1980-1990 Asia lost its rainforests at a rate of 11%, which is much higher than the seven-odd percent rate in Africa and Latin America (Fig. 2). Over the three decades from 1960 to 1990 Asia lost about one-third of its rainforests. Although the Philippines and Thailand were once the main exporters of tropical timber, they are now wood importers because of their excessive logging.

Fig.1 Tropical Forest Hotspots

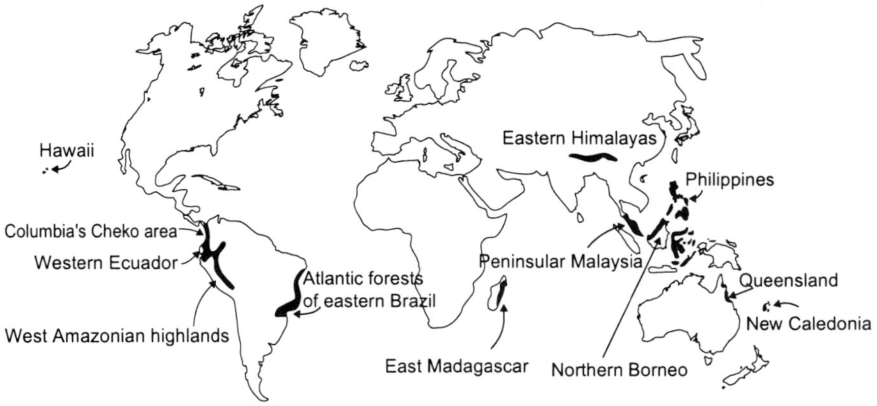

Source: Myers, N., *Threatened Biotas: "Hotspots" in Tropical Forests*, 1988.

Fig.2 Rate of Tropical Forest Loss, 1960-1990

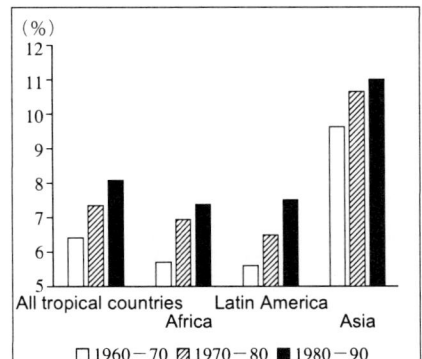

Source: World Resources Institute, *World Resources 1996-1997* (1996).
Originally from: K.D. Singh and Antonio Marzoli, "Deforestation Trends in the Tropics: A Time Series Analysis," paper presented at the World Wildlife Fund Conference on the Potential Impact of Climate Change on Tropical Forests, San Juan, Puerto Rico, April 1995, pp. 8-9.

Fig.3 Ten Top Mangrove Countries

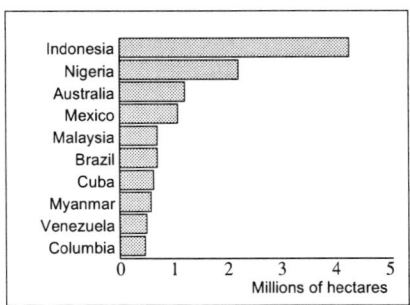

Note: Included are only countries with data on the sizes of their mangrove areas. There was only one datum each for Australia and Myanmar. The current mangrove areas for the other countries were obtained by averaging the highest and lowest estimates. Data are from different years.
Source: Same as Fig.2.
Originally from: World Conservation Monitoring Centre, *Biodiversity Data Sourcebook*, World Conservation Press, Cambridge, U.K., 1994, pp. 74-98.

Coastal ecosystems like mangrove forests and coral reefs face the same grim situation. While 20-30% of coastal areas in Africa, Latin America, and Oceania are under serious potential threat from development such as port and harbor construction, the figure for Asia is 52%.

Mangrove forests line one-fourth of the world's tropical coasts, but they are concentrated in just a few countries. Asian countries in the top 10 are Indonesia, Malaysia, and Myanmar (Fig. 3). Mangroves from the Indian Ocean to the West Pacific are in crisis because of cutting for fuelwood, conversion for shrimp farming, and other purposes. The Philippines has already lost 70% of its old-growth mangrove forests.

Coral reefs are one of the most diverse ecosystems, and are said to harbor one-fourth of all marine species. Fifteen percent of the world's coral reefs are concentrated into the region from the Indian Ocean to the West Pacific, but this coral has a grim future owing to red soil runoff, the digging of coral for building material, catching of fish using dynamite and cyanide, and other causes. Coral bleaching, a phenomenon observed worldwide in recent years, is said to occur because of a 1°C rise in sea water temperature, and it is likely that coral will face a still bleaker future as global warming proceeds.

This damage to terrestrial and coastal ecosystems is pushing their species toward extinction. Proportions of globally endangered species in Asian countries are, for mammals, over 10% in almost all countries including China and Indonesia. Birds have high percentages in the Philippines, while reptiles and amphibians are high in Japan. Japan has by far the highest percentage among Asian countries of endangered higher plant species (Table 1).

Japan was among the first Asian countries to initiate research on endangered plant species, and has prepared a Red Data Book on higher plants, but it is also a fact that in this same country wild plant habitat has been destroyed by rapid economic growth, and that orchids and other rare species have gone extinct because of commercial exploitation.

Characteristic of threats to wild animals in Asia is poaching to obtain materials such as rhinoceros horn, tiger bones, bear gallbladders, and the musk pouches of musk deer, which are used in Chinese medicines, perfume, and other products. Depite efforts toward regulating international trade in these items and toward developing substitutes, the increasing affluence of Asians gives rise to concerns that demand for Chinese remedies, perfume, and other such products will continue to grow.

One of the most effective ways to conserve biodiversity is to establish protected areas. At the World Congress on National Parks and Protected Areas in 1992 the World Conservation Union (IUCN) set forth a goal of making 10% of the Earth's land into protected areas, but the only places to have accomplished this so far are Japan and Hong Kong.

(YOSHIDA Masato)

Table 1 Globally Endangered Species, 1993

	1 China	2 Indonesia	3 Japan	4 S. Korea	5 Malaysia	6 Philippines	7 Thailand	8 Taiwan
A. Mammals (total species)	394	436	132	49	286	153	265	213
Endangered species	42	57	17	6	20	22	22	25
(%)	11%	13%	13%	12%	7%	14%	8%	12%
B. Birds (total species)	1,244	1,531	583	372	736	556	915	761
Endangered species	86	104	31	19	31	86	44	45
(%)	7%	7%	5%	5%	4%	15%	5%	6%
C. Reptiles (total species)	340	511	66	25	268	190	298	180
Endangered species	8	16	10	0	10	8	11	8
(%)	2%	3%	15%	0%	4%	4%	4%	4%
D. Amphibians (total species)	263	270	52	14	158	63	107	80
Endangered species	1	0	11	0	0	2	0	1
(%)	0.4%	0%	21%	0%	0%	3%	0%	1%
E. Fish (total species)	686	-	186	130	449	-	>600	-
Endangered species	16	65	10	0	4	21	11	2
(%)	2%		5%	0%	1%			
F. Higher plants (total species)	30,000	27,500	4,700	2,898	15,000	8,000	11,000	>7,000
Endangered species	343	281	704	69	510	371	382	350
(%)	1%	1%	15%	2%	3%	5%	3%	

Source: World Conservation Monitoring Centre.

[13] Advancing Urbanization

Urbanization is on the march throughout the world as never before. According to UN statistics,[1] although the urban population in 1950 was only about 700 million, it will probably attain 5 billion in 2025. The level of urbanization (i.e., the percent of total population in urban settlements) has consistently risen, from 29.3% in 1950 to 43.1% in 1990, and in 2025 it is projected to reach 61.1%. Rapid Asian urbanization is pushing up the overall worldwide rate of urbanization. Asia's current urbanization level is relatively low at an average of about 30%, but as we move into the 20th century it will rise quickly (Fig. 1).

Asian countries with high urbanization levels of 60% or more are presently Japan and the NIEs. What they all have in common is the experience of rapid economic growth and industrialization. There is a clear connection between economic level and urbanization in which generally countries that are highly developed economically also have high urban population percentages. Malaysia and the Philippines have started economic growth and urbanization in an attempt to emulate Japan and South Korea. The urban population levels in the populous Asian countries of China, India, and Indonesia in 1994 were low at 29.4%, 26.5%, and 34.4%, respectively, but henceforth they will probably surge in conjunction with economic growth.

Although urbanization is also caused by natural population increase, of special note in Asia is social increase, or the influx of people from rural to urban areas. Social increase is a phenomenon generally observed as urbanization proceeds, but in Asia it is especially pronounced (Fig. 2).

Asia's pace of urbanization is far faster than that in developed Western countries. For example, Great Britain, France, and the United States required 70-80 years, 100 years, and 60 years, respectively, to raise their urbanization percentages from 20 to 50%, and a further increase from 50 to 70% took 35, 28, and 40 years, respectively. In Japan, however, the increase from 20 to 50% took 31 years, and that from 50 to 70% took 15 years, or about half the time. Urbanization occurred even faster in South Korea, which required 30 years to go from 20 to 50%, and a mere decade from 50 to 70%.

---

[1]   Because the definition of "city" differs from one country to another, there are hardly any data allowing urban population comparisons among countries. What is more, there are considerable differences between the urban population statistics of the UN and the World Bank. Here I have used UN data on recent urbanization: *World Urbanization Prospects: The 1994 Revision* (ST/ESA/SER.A/150)

Fig.1 Urbanization, 1950-2025

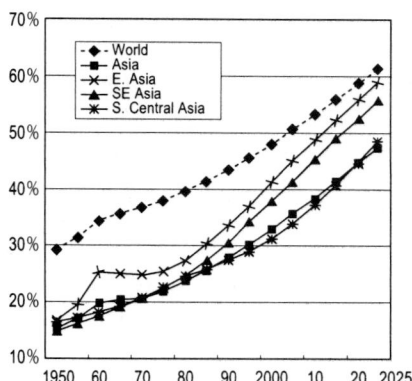

Source: Prepared from United Nations, *World Urbanization Prospects: The 1994 Revision* (1995).

Fig.2 Average Annual Rates of Increase in Total and Urban Populations, 1970-94

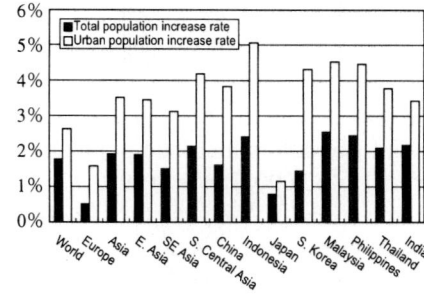

Source: Same as Fig.1.

Fig.3 The 20 Largest Urban Agglomerations, 1990

(thousands)

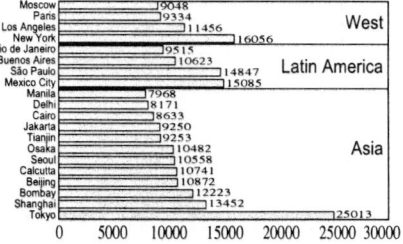

Note: The graph includes Cairo in Asia.
Source: Same as Fig.1.

Table 1 Population and Urbanization in Asia

	Year	World	Asia	China	Indonesia	Japan	S. Korea	Malaysia	Philippines	Thailand
A. Population	1950	251,975	140,273	55,476	7,954	8,363	2,036	611	2,099	2,001
(10,000s)	1970	369,714	214,749	83,068	11,067	10,433	3,192	1,085	3,754	3,575
	1994	562,963	340,344	120,884	19,461	12,482	4,456	1,970	6,619	5,818
	2010	703,229	426,395	138,847	23,960	12,715	5,076	2,624	8,816	6,713
B. Avg. annual	1950-70	1.9%	2.2%	2.0%	1.7%	1.1%	2.3%	2.9%	2.9%	2.9%
growth rate of	1970-94	1.8%	1.9%	1.6%	2.4%	0.7%	1.4%	2.5%	2.4%	2.1%
population (%)	1994-2010	1.4%	1.4%	0.9%	1.3%	0.1%	0.8%	1.8%	1.8%	0.9%
C. Urban population	1950	73,785	23,579	6,102	986	4,207	435	124	570	210
(10,000s)	1970	135,279	50,295	14,495	2,053	7,430	1,300	363	1,238	475
	1994	252,051	115,933	35,560	6,702	9,676	3,565	1,042	3,518	1,149
	2010	370,707	189,017	59,677	11,904	10,250	4,639	1,691	5,873	1,840
D. Rural population	1950	178,190	116,694	49,374	6,968	4,156	1,601	487	1,529	1,791
(10,000s)	1970	234,436	164,454	68,572	9,014	3,004	1,893	722	2,516	3,100
	1994	310,912	224,411	85,324	12,759	2,805	892	927	3,101	4,670
	2010	332,522	237,378	79,171	12,056	2,465	438	933	2,943	4,873
E. Percentage of	1950	29.3%	16.8%	11.0%	12.4%	50.3%	21.4%	20.4%	27.1%	10.5%
population residing	1970	36.6%	23.4%	17.5%	18.6%	71.2%	40.7%	33.5%	33.0%	13.3%
in urban areas	1994	44.8%	34.1%	29.4%	34.4%	77.5%	80.0%	52.9%	53.1%	19.7%
	2010	52.7%	44.3%	43.0%	49.7%	80.6%	91.4%	64.4%	66.6%	27.4%
F. Avg. annual	1950-70	3.1%	3.9%	4.4%	3.7%	2.9%	5.6%	5.5%	4.0%	4.2%
growth of urban	1970-94	2.6%	3.5%	3.8%	5.1%	1.1%	4.3%	4.5%	4.4%	3.7%
population (%)	1994-2010	2.4%	3.1%	3.3%	3.7%	0.4%	1.7%	3.1%	3.3%	3.0%
G. Avg. annual	1950-70	1.4%	1.7%	1.7%	1.3%	-1.6%	0.8%	2.0%	2.5%	2.8%
growth of rural	1970-94	1.2%	1.3%	0.9%	1.5%	-0.3%	-3.1%	1.0%	0.9%	1.7%
population (%)	1994-2010	0.4%	0.4%	-0.5%	-0.4%	-0.8%	-4.3%	0.0%	-0.3%	0.3%

Note: This table uses "Asia" for the source minus the figures for "Western Asia." Figures for 2010 are forecasts. China includes Taiwan.

Source: Prepared from United Nations, *World Urbanization Prospects: The 1994 Revision* (1995).

In Asia the formation of megacities is also very pronounced. Cities defined by the UN as "megacities," i.e., those with populations of at least 8 million, are growing in number especially in Asia. In 1950 the only megacities were New York and London, which had populations of only 102,339,000 and 8,733,000. But as of 1990, 12 of the world's biggest 20 cities were in Asia, and the largest of them — Tokyo — had a population of 25 million (Fig. 3). Predictions say that in 2015 the world will have 33 megacities, 22 of which will be in Asia.

In Asia's megacities the rate of population increase has long been very high. Of the major Asian cities with populations of 5 million or more, Dacca, Karachi, Jakarta, Delhi, Manila, Bombay, Seoul, and Bangkok have had annual growth rates of at least 3% over the more than 40 years since 1950. It is said that generally when population grows at over 3% annually for at least a few years, infrastructure cannot be built fast enough. When Japan's large cities experienced serious problems from the latter half of the 1960s into the first half of the 1970s, their population growth rates were in fact over 3%, and they could not build water, sewerage, roads, schools, and other infrastructure fast enough. The phenomena now appearing in Asia's large cities might well be considered a large-scale, long-term reenactment of Japan's urbanization, and one that is unprecedented worldwide. Having achieved such rapid growth, Asia's megacities now suffer from the lack of sewage treatment facilities and other infrastructure, and from worsening air and water pollution.

(OSHIMA Ken'ichi)

[14] Asia's Growing Energy Use

Owing to a huge population and rapid economic growth, Asia's primary energy consumption is growing quickly, increasing about 2.8-fold during the approximately 20 years from 1971 to 1993, going from 680 million TOE (tons oil equivalent) to 1,870 million TOE (Table 1). The average annual rate of increase over these years was 5.8% (excluding Japan), which was far higher than the world average of 2.3% and the OECD average of 1.5%. Some Asian countries, including South Korea, Indonesia, Taiwan, Thailand, Malaysia, and Singapore, increased at rates of over 8%, thereby raising Asia's share of world primary energy consumption from 13.9% in 1971 to 23.6% in 1993 (Fig. 1).

A look at Asia's energy consumption in terms of demand shows that the increased supply of energy was induced mainly by higher industrial energy demand. Energy consumption by the industrial sector in Asia is high, accounting for 56.8%, and that sector is also responsible for 50% of the increase in energy demand between 1980 and 1993, making

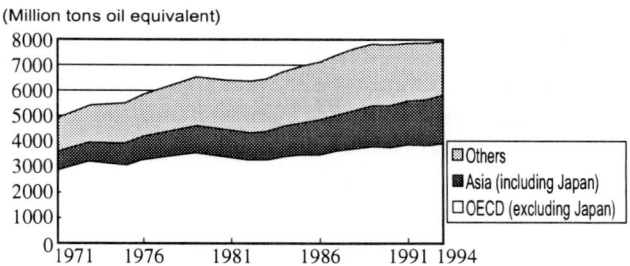

Fig.1 World Primary Energy Supply

(Million tons oil equivalent)

Sources: Prepared from OECD/IEA, *Energy Statistics and Balances of Non-OECD Countries 1992-1993*, 1995.
OECD/IEA, *Energy Statistics of OECD Countries 1992-1993*, 1995.

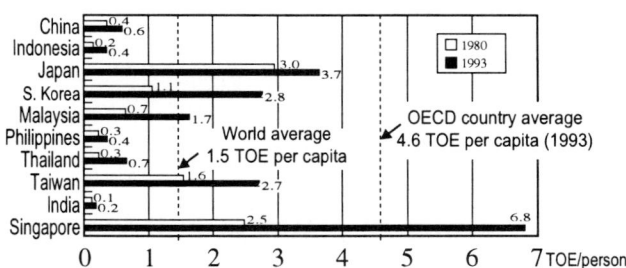

Fig.2 Per Capita Energy Consumption

Sources: Same as Table 1.

Table 1 *Economic Growth Rate and Average Annual Growth Rates of Primary Energy Consumption and Generating Capacity, 1971-93 (%)*

	GDP	Primary energy consumption	Electricity production
World	3.1	2.3	4.6
OECD	2.8	1.5	3.4
Asia*		5.8	6.5
China	8.1	5.3	8.5
Indonesia	6.4	9.9	14.5
Japan	3.9	2.4	4.0
S. Korea	8.7	9.6	12.6
Malaysia	7.2	8.8	10.7
Philippines	3.2	4.5	5.1
Thailand	7.6	8.7	12.2
Taiwan	11.0	8.0	9.1
India	4.3	5.7	7.9
Singapore	7.8	8.8	9.5

Note: Asia was determined to be those countries categorized in OECD/IEA's *Energy Statistics of Non-OECD Countries* as "Asia" plus China and Japan, for a total of 43 countries/regions.

Sources: OECD/IEA, *Energy Statistics and Balances of OECD Countries 1992-93*.
OECD/IEA, *Energy Statistics and Balances of Non-OECD Countries 1992-93*.
Directorate-General of Budget, Accounting and Statistics, Executive Yuan, Republic of China, *Statistical Yearbook of the Republic of China 1995*.
World Bank, *World Tables*, 1995.

it safe to say that industrialization is the primary factor inducing higher energy supply.

Developed countries indicate concern over this spiraling Asian energy consumption from the perspective of conserving the global environment, and especially preventing global warming, but the developed countries still monopolize the world's energy resources. People in the 10 biggest energy consuming Asian countries (listed in Table 1) consume only 0.68 TOE per capita (0.53 TOE if Japan is excluded), which is below the world average, and a mere one-seventh of the OECD average of 4.6 TOE per capita (Table 2).

But it is also a fact that in the 1980s and 1990s some Asian countries, mainly among the NIEs, are making the transition to societies that are as energy-intensive as developed countries. In 1993 per capita energy consumption in South Korea and Taiwan had already approached Japan's 1980 level, while Singapore had far exceeded Japan and now ranks with the U.S. among the world's biggest energy consumers.

In terms of energy supply China and India, which account for the greater part of Asia's energy consumption, both rely chiefly on coal, whose proportions of their primary energy supply are 76.5% and 59.6%, respectively. As coal is a relatively dirty energy source, and because these countries take hardly any environmental measures such as installing desulfurization equipment or precipitators, the resulting environmental problems are serious.

Another matter deserving attention is the increasingly active nuclear power development in Asia as an environmental measure. Behind this is the tight electricity supply engendered by the sharply increasing demand for electric power. Asian countries with nuclear power programs are Japan, South Korea, Taiwan, China, India, and Pakistan, with Indonesia and Thailand trying to join their ranks. South Korea, Taiwan, and Japan are already among the world's major nuclear power nations, depending on nuclear for 40%, 32%, and 28% of their total electricity supplies, respectively (1993). And while France and other Western countries have abandoned their fast breeder reactor programs, China has plans for building an experimental reactor by 2000 and generating power with a commercial reactor in the early 21st century. Even Indonesia, which as yet does not rely on nuclear, passed a law in February 1997 that is to facilitate the introduction of nuclear power.

Since the 1990s transferring technologies for pollution control and energy conservation has become a mainstay of Japan's international cooperation. While this in itself is laudable, even if technical transfer allows other Asian countries to attain high energy efficiencies equal to that of Japan, it will not constitute a fundamental solution if Asian countries merely copy the energy-intensive societies of developed nations. The real challenge for developing Asian countries is to avoid creating energy-wasting societies like those of developed countries and instead build energy supply/demand structures that impose little burden on the environment. Their tasks include achieving efficient energy use, and the development and deployment of renewables like photovoltaic and wind power, which will require a greater degree of international cooperation.

(OSHIMA Ken'ichi)

Table 2 Energy Supply and Demand in Asia, 1993

	1 Asia total	2 China	3 Indonesia	4 Japan	5 S. Korea	6 Malaysia	7 Philippines	8 Thailand	9 Taiwan
A. Primary energy (10,000 TOE)									
Domestic production	134,065	74,093	14,426	8,387	2,001	5,440	675	1,718	1,054
Imports	84,549	3,449	1,530	39,015	12,624	1,269	1,675	2,626	4,987
Exports	-26,282	-3,503	-8,934	-828	-1,442	-3,523	-93	-106	-48
Inventory change, etc.	-4,992	-945	-115	-830	-846	-11	16	-178	-238
Total domestic primary energy supply	187,340	73,094	6,906	45,744	12,338	3,176	2,272	4,060	5,755
B. Primary energy supply (10,000 TOE)									
Oil	73,092	14,398	4,382	25,587	7,637	1,864	1,605	2,659	2,896
Natural gas	14,181	1,418	1,870	4,766	569	1,136	0	840	288
Coal	86,197	55,885	490	7,683	2,564	134	125	524	1,616
Nuclear	9,105	42	0	6,496	1,515	0	0	0	897
Hydro	3,659	1,306	58	841	52	42	42	32	58
Others	1,107	45	107	371	0	0	501	5	0
Combustible Renewables and Waste	21,759	5,409	3,570		74	231	955	1,773	0
C. Final energy consumption (10,000 TOE)	133,946	55,902	4,403	31,642	9,764	1,878	1,322	2,856	3,979
Industry sector	76,070	37,076	1,741	14,789	5,079	864	438	993	2,287
Transport sector	26,192	5,520	1,386	8,419	2,030	656	582	1,424	989
Commerce and public sector	5,986	1,442	99	3,043	346	72	94	189	207
Residential sector	20,955	9,859	892	4,499	2,303	138	166	211	354
Other sectors	2,143	1,242	148	6	0	7	10	5	55
Non-energy use	2,600	763	137	886	6	142	32	34	87
D. Electricity Production (100 million kWh)	27,300	8,395	501	8,990	1,444	356	274	634	1,071
Oil	4,347	728	237	1,799	349	133	147	182	264
Coal	11,900	6,106	117	1,679	309	42	21	135	373
Gas	3,037	27	68	1,929	145	132	0	280	23
Hydro	4,254	1,518	67	978	60	49	49	37	67
Nuclear	3,494	16	0	2,493	581	0	0	0	344
Geothermal	93	0	12	18	0	0	57	0	0
Others	175	0	0	95	773	0	0	0	0
E. Electricity demand (100 million kWh)	22,924	6,931	416	7,966	1,269	309	218	563	924
Industrial sector	14,782	5,668	219	4,369	757	165	94	218	486
Commerce and public sector	3,035	425	35	1,341	260	83	49	220	165
Transport sector	415	132	0	209	13	0	0	0	5
Residential sector	4,311	705	134	2,046	239	53	64	119	218
Other sectors	381	0	27	0	0	8	11	6	51

Notes: 1: The total for Asia was determined to be those countries categorized in OECD/IEA's *Energy Statistics of Non-OECD Countries* as "Asia" plus China and Japan, for a total of 43 countries/regions. B: Total primary domestic energy supply does not include combustible renewables and wastes.

Sources: OECD/IEA, *Energy Statistics and Balances of OECD Countries 1992-93.*
OECD/IEA, *Energy Statistics and Balances of Non-OECD Countries 1992-93.*

[15] Air Quality in Major Asian Cities

Any visitor to Bangkok, Manila, Jakarta, or other major Asian cities will notice how dirty the air is. But mainly what we sense are dust and the odor arising from hydrocarbons (except for China's Guiyang and other cities with serious SO_2 air pollution), which are, strictly speaking, not air pollutants. The dust that gets into our eyes and noses and causes physical discomfort consists of particles at least several tens of micrometers in size that will eventually settle to the ground, unlike the suspended particulates that enter the inner recesses of our lungs (under about 10 micrometers in size). There is some research suggesting that hydrocarbons affect brain functions and other physical processes, but ordinarily they are in the spotlight merely as substances that, with nitrogen oxides, trigger the formation of ozone. It is therefore important to note the incongruity between what we feel is pollution, and what is scientifically defined as such.

But it is also very difficult to ascertain the state of "scientifically defined air pollution." Presented here are data on the average annual concentrations of six air pollutants in major Asian cities, but only the data from Tokyo, Seoul, and Taipei are average annual concentrations in the true sense of the word (Table 2). Obtaining the average annual concentration requires that measurements be made every hour or every day continuously over a year, but it is doubtful that such measurements are made in cities other than the above three. The author has personally observed that, at least in Bangkok, Manila, and Jakarta, not a few monitoring stations perform no measurements owing to equipment breakdowns, inadequate instrumentation, insufficient funding, and other problems.

Further, even when measured values are reported, one must in not a few cases doubt their accuracy. As air pollution measurements involve minute analyses on the order of one one-billionth of a unit, the slightest errors in measurement methods or lapses in instrument maintenance and care mean that results often differ from true values by as many as two orders of magnitude. Such deviant values are excluded from readings as errors, but when there are too many of them the data are sometimes massaged.

Therefore in determining the state of air pollution in major Asian cities, making a rough estimate is more important than exact data measurements.

A look at air pollution by pollutant types according to five levels (Table 1) shows that, except for Seoul and Kuala Lumpur, pollution by suspended particulates is very serious. The main sources of these particulates are motor vehicle exhaust, and the dust raised by vehicle traffic and building construction work (Table 3).

Many cities also have grave NO_2 air pollution. Deserving special note here is that Beijing, Shanghai, Seoul, and other cities had lower NO_2 levels prior to the 1990s, but had more serious SO_2 pollution then. This is likely because the main source of pollutants in those cities has shifted from factory and household-service use of high-sulfur fuel oil and coal to motor vehicles.

Finally, because ozone is formed in photochemical reactions involving non-methane hydrocarbons (contained in large amounts in motor vehicle exhaust) and nitrogen oxides, cities with heavy NO_2 pollution generally also have serious ozone pollution. But because cities also have large amounts of reducing substances (mainly NO), it is often the case that the areas around cities have worse ozone pollution than urban areas themselves. For example, the rural areas about 50 km from Jakarta are known for their heavy ozone pollution.

Table 1 Air Pollution in Major Asian Cities

City	Beijing	Shanghai	Guiyang	Jakarta	Tokyo	Seoul	Kuala Lumpur	Manila	Bangkok	Taipei
TSP	E	D	E	D	C	B	B	E	E	D
SO_2	C	B	E	A	A	B	B	B	-	B
CO	-	-	-	B*	A	A	C	-	B	B
NO_2	D	C	B	B	C	C	C	-	(C)	B
O_3	(C)	-	-	(C)	C	B	C	-	-	-
Pb	-	-	-	C	-	A	-	D	A*	-

Assessment criteria

Atmospheric concentrations of pollutants are relative to WHO standards.

A: under half; B: within standard; C: under twice standard; D: under three times standard; E: over three times standard. Pb levels are roadside, all others are ambient air. Asterisks indicate that reported readings are much lower than suggested by actual conditions. Parentheses indicate that assessments are estimates based on sources other than those in Table 2.

Examining air pollutant emissions in cities by sector makes it possible to generally determine the main sources of each pollutant (Table 4). However, because people arrive at such calculated values by assuming emission intensity (the amounts of pollutants emitted per unit of fuel) and the number of units (as far as can be ascertained), it is best not to attach too much significance to such values. For example, a study and review by WHO of the calculated values for air pollutant emissions in Manila (Table 4-8) showed that sulfur oxides from the industrial and commercial sectors, suspended particulates from diesel-powered vehicles, and carbon monoxide from diesel vehicles were assessed at five times, 10 times, and 0.5 times their actual levels, respectively.

(MIZUTANI Yoichi)

Table 2 Air Pollution in Major Asian Cities

	1	2	3	4	5	6	7	8	9	10
City	Beijing (1995)	Shanghai (1995)	Guiyang (1995)	Jakarta (1993/95)	Tokyo (1995)	Seoul (1995)	Kuala Lumpur (1992/93)	Manila (1991/92)	Bangkok (1994/95)	Taipei (1993)
A. Ambient air (residential areas)										
A1 Total suspended particulates ($\mu g/m^3$)	370	246	330	82.2-247.6	48[SPM]	85	29.2-81.7 (PM$_{10}$19.8-55.4)		<550 (Max.24-h)	240
A2 SO$_2$ ($\mu g/m^3$)	94	53	424	0.0006-0.0065ppm	0.007ppm	0.017ppm	0.021-0.0335ppm (Max.24-h)			0.017ppm
A3 CO (mg/m^3)					0.8ppm	1.3ppm	3.55-10.53ppm (Max. 8-h)		<17 (Max.1-h)	4.94ppm
A4 NO$_2$ ($\mu g/m^3$)	124	75	53	0.0054-0.0130ppm	0.031ppm	0.032ppm	0.0072-0.168ppm (Max.1-h)			0.025ppm
A5 Ozone ($\mu g/m^3$)					0.025ppm	0.013ppm	0.0558-0.1187ppm (Max. 8-h)			
A6 Pb ($\mu g/m^3$)				0.2569-0.5124		0.1844			<0.4* (Max.1-h)	
B. Roadside air										
B1 Total suspended particulates($\mu g/m^3$)				342-441	0.065ppm[SPM]			290-497 (PM$_{10}$134-224)	64-1130 (PM$_{10}$10-314) (24-h)	
B2 SO$_2$ ($\mu g/m^3$)				0.006-0.019ppm	0.011ppm			0.02ppm		
B3 CO (mg/m^3)				2.589ppm*	1.9ppm			20.6 (Max.1-h)	0-45.2 (1-h)	
B4 NO$_2$ ($\mu g/m^3$)				0.046ppm	0.042ppm			0.01		
B5 Ozone ($\mu g/m^3$)								<0.01 (Max.1-h)		
B6 Pb ($\mu g/m^3$)				0.91-1.14				1.0-2.3	0-0.7* (24-h)	

Notes: Atmospheric concentrations are annual average values when there are multiple monitoring stations; when average is unknown, lowest and highest values are given.

Monitoring stations of unknown types are all classified as ambient air (residential area) stations.

A-4: 7 sites, 1995. B1-4, B2-4, B6-4: 4 or 5 sites, 1993. B3-4, B4-4: Jl. Thamrin, 1995.

A1-7, A2-7, A3-7, A4-7, A5-7: 3 or 5 sites.

A1-9, A3-9, A6-9: 1995. B1-9, B3-9, B6-9: 1994.

A1-10, A2-10, A3-10, A4-10: Average of 11 stations.

Asterisks indicate that reported readings are much lower than suggested by actual conditions.

Sources: 1-3: Data for Beijing, Shanghai, and Guiyang are from *China Environment Yearbook 1996*.

4: 1993 data for Jakarta are from Kantor Pengkajian Perkotaan dan Lingkungan DKI Jakarta, *Laporan Lingkungan Jakarta '93-1994 Udara dan Kebisingan* (1994); data for 1995 are from the same source for 1995-1996 (1996).

5: Data for Tokyo are from "Overview of Air Pollution Measurement Results for 1995" (August 1996) by the Metropolitan Tokyo Bureau of Environmental Protection.

6. Seoul data are from the *1996 Environment White Paper* by South Korea's Environment Ministry.

7. Data for Kuala Lumpur are from Department of Environment, Ministry of Science, Technology and Environment documents.

8. Manila data are from the Asian Development Bank's *Final Report — Vehicular Emissions Control Planning Project in Metro Manila* (1993).

9. Bangkok data are from Ministry of Science, Technology and Environment documents.

10. Taipei data are from *Republic of China Taiwan Region Environmental Data 1983* by the Administrative Yuan, Environmental Protection Administration.

Table 3 Air Quality Standards in Asian Countries[1]

	1 China[2]	2 Indonesia	3 Japan	4 S. Korea	5 Malaysia	6 Philippines	7 Thailand	8 Taiwan	9 WHO[3]
Total suspended particulates (μg/m³) (24-h mean)	150/300/500 (PM$_{10}$ 50/150/250)	260	100 [SPM]	300	260 (PM$_{10}$ 150)	180	330 (PM$_{10}$ 120)	250 (PM$_{10}$ 125)	150-230
(8-h mean)	-/420/680								
(1-h)			200 [SPM]			250			
(annual mean)				150	90 (PM$_{10}$ 50)		100 (PM$_{10}$ 50)	130 (PM$_{10}$ 65)	60-90
SO$_2$ (μg/m³) (1-h)		260(0.1ppm)	260(0.1ppm)	429(0.15ppm)	372(0.13ppm)	850	750(0.3ppm)	715(0.25ppm)	
(24-h mean)	50/150/250		114(0.04ppm)		114(0.04ppm)	369	300(0.12ppm)	286(0.1ppm)	100-150
(annual mean)	20/60/100			143(0.05ppm)			100(0.04ppm)	85.8(0.03ppm)	40-60
CO (mg/m³) (1-h)					34(30ppm)	35	34.4(30ppm)	40(35ppm)	30
(8-h mean)		22.6(20ppm)	22.6(20ppm)	22.9(20ppm)	10(9ppm)	10	10.3(9ppm)	10.3(9ppm)	10
(others)	4/4/6(24-h)		11.45(24-h)	9.2(8ppm, monthly mean)					
NO$_2$ (μg/m³) (1-h)				282(0.15ppm)	201(0.17ppm)	190	320(0.17ppm)	470(0.25ppm)	400
(24-h mean)	50/100/150[NO$_x$]	92.5[NO$_x$]	75-113					150	
(annual mean)		(0.05ppm)↑	(0.04-0.06ppm)↑	92.5(0.05ppm)				94(0.05ppm)	
Ozone (μg/m³) (1-h)	120/160/200	160(0.08ppm)	120(0.06ppm)	200(0.1ppm)	200(0.1ppm)	120	200(0.1ppm)	240(0.12ppm)	150-200
(8-h mean)					120(0.06ppm)			120(0.06ppm)	100-120
(annual mean)			40(0.02ppm)						
Pb (μg/m³) (24-h mean)		60					10	10 (monthly mean)	
(annual mean)				1.5 (3-month mean)	1.5 (3-month mean)		1.5 (monthly mean)	1.5 (quarterly mean)	0.5-1

Notes: 1. Unless especially prescribed by a country's environmental quality standards, μg/m³-ppm conversion coefficients usually use the following values. 1 ppm SO$_2$ = 2,860 μg/m³. 1 ppm CO = 1.145 μg/m³. 1 ppm NO$_2$ = 1,880 μg/m³. 1 ppm O$_3$ = 2,000 μg/m³.
2. Class I/Class II/Class III. Class I: Tourist, historical, and conservation areas. Class II: Residential urban areas and rural areas. Class III: Industrial areas and heavy traffic areas.
3. WHO standards are those established by WHO's Regional Office for Europe in 1987 (World Health Organization, Regional Office for Europe, *WHO 1987 Air Quality Guidelines for Europe*, WHO Regional Publications, European Series No. 23, Copenhagen).

Source: Table created by author using country data based on UNEP and WHO, *Urban Air Pollution in Megacities of the World* (Blackwell, Oxford U.K., 1992).

Table 4 Air Pollutant Emissions in Major Asian Cities

City (tons of emissions per year)	1 Beijing (1995)	2 Shanghai (1995)	3 Guiyang (1995)	4 Jakarta (1991)	5 Tokyo (1990/95)	6 Seoul (1994)	7 Kuala Lumpur (1992)	8 Manila (1987/88)	9 Bangkok (1994)	10 Taiwan overall (1993)
A. Main sources of TSP emissions	278,706	185,643	59,905	6,762.5	('90) 11,830	17,245	1,536	69,100		740,224 (PM$_{10}$ 351,647)
A.1 Industrial processes	124,471	133,343	33,305	1,076.8	1,370	652	346	6,000		64.40%(46.32%)
A.2 Transport/vehicles				3,252.0	9,870	14,420	1,190	57,100	31,691	6.74%(14.34%)
A.3 Household-service (residences, commerce)				2,433.8	480	2,157		6,000		
A.4 Others				619.5	110	16				
B. Main sources of SO$_x$ emissions	382,925	488,564	160,282	28,182.1	('95) 12,900	40,127	1,670	148,400		463,410
B.1 Industrial processes	214,899	381,454	111,007	17,687.7	6,100	7,788	641	31,000		18.16%
B.2 Transport/vehicles				7,476.0	6,700	5,107	1,029	34,400		5.22%
B.3 Household-service (residences, commerce)				3,018.5	100	27,177				
B.4 Others				56.4		55		83,000		
C. Main sources of CO emissions				374,418.4	('90) 200,300	255,154	290,407	557,000	2,107,379	
C.1 Industrial processes				378.2	28,000	477		2,000	2.92%	
C.2 Transport/vehicles				373,662.0	172,300	209,588	290,407	554,000	2,313,556	82.28%
C.3 Household-service (residences, commerce)				378.2		44,834				
C.4 Others				3,782.0		255		1,000		
D. Main sources of NO$_x$ emissions				20,734.1	('95) 67,600	112,745	13,620	119,000	414,499	
D.1 Industrial processes				3,333.4	9,900	3,533	102	1,000		15.00%
D.2 Transport/vehicles				15,388.0	48,300	93,027	13,518	109,000	134,249	48.35%
D.3 Household-service (residences, commerce)				2,012.6	9,400	12,759				
D.4 Others				230.6		3,426		9,000		

Notes: A2, B2, C2, D2-6,7,10: Motor vehicles only (including two-wheeled vehicles).
   A2-8: Diesel vehicles: 54,000 t/year, gasoline vehicles: 3,000 t/year, aircraft: 100 t/year.
   B2-8: Diesel vehicles: 31,000 t/year, gasoline vehicles: 3,000 t/year, aircraft: 400 t/year.
   C2-8: Diesel vehicles: 27,000 t/year, gasoline vehicles: 514,000 t/year, aircraft: 13,000t/year.
   D2-8: Diesel vehicles: 76,000 t/year, gasoline vehicles:29,000 t/year, aircraft: 4,000t/year.
   A2-10: Road traffic only, 6.64% (13.72%).
   A4-10: Scatter particles (dust raised by vehicular traffic, construction work, etc.), 19.75% (24.61%); fuel combustion, 4.28% (5.15%); non-fuel combustion, 4.83% (10.03%).
   B2-10: Road traffic only, 5.16%. B4-10: Fuel combustion, 76.00%. Non-fuel combustion, 0.053%.
   C4-10: Fuel combustion, 0.87%. Non-fuel combustion, 13.93%. D2-10: Road traffic only, 47%. D4-10: Fuel combustion, 36.20%. Non-fuel combustion, 0.16%.

Sources: 1-3: For Beijing, Shanghai, and Guiyang, China Environmental Yearbook 1996.
   4: Data for Jakarta are from Environmental Impact Management Agency (BAPEDAL) documents.
   5: Tokyo data from the Metropolitan Tokyo Bureau of Environmental Protection.
   6: Seoul data from Environment Ministry's 1996 Environment White Paper and Environment Office Atmospheric Protection Bureau documents.
   7: Kuala Lumpur data from Department of Environment, Ministry of Science, Technology and Environment documents.
   8: Manila data from Environmental Management Bureau, Department of Environment and Natural Resources, 1988 Metro Manila Air Pollution Emission Inventory for 1987 (1990).
   9: Motor vehicle emissions for Bangkok in 1994 based on Ministry of Science, Technology and Environment documents.
   10: Taiwan data are from Republic of China Taiwan Region Environmental Data 1983 by the Administrative Yuan, Environmental Protection Administration.

[16] Tightening Water Supplies

Agriculture still accounts for a large proportion of water demand in Asia. Not only does irrigation substantially raise the productivity of farmland, many of the high-yield crop varieties developed in recent years cannot be grown without irrigation. This makes it likely that as the food supply tightens in the future, so will supplies of water.

Many countries try to meet the growing demand for water by building dams or pumping more groundwater. Under construction on China's Chang Jiang (Yangtze) River is the huge Three Gorges dam, which will be 185 m high, have a reservoir of 1,084 square km, and impound 39.3 billion tons of water. On the upper reaches of the Mekong River, known in China as Lancang Jiang, construction is in progress on a series of 14 dams. The first of these, the Manwan Dam (132 m high, 23.9 square km reservoir, 920 million tons impounded), was completed in 1996. Most development on the upper reaches of the Mekong, however, is for hydropower.

Dam development causes a number of problems, first of which is the relocation of people living in areas that will be inundated. The Three Gorges Dam, for example, will displace over 1.1 million people. Such relocation in turn causes problems of increased stress on the displaced people and on the communities that take them in, as well as the burden of compensation.

Dam construction also strains public finances. What is more, cost overruns in dam construction are more the rule than the exception.

Building dams engenders considerable change in rivers' aquatic environments, and sometimes triggers changes in their biota. Especially in the tropics there are instances in which stagnation of river flows allowed mosquitoes, schistosomes, and filariae to breed, thereby helping the spread of malaria and other diseases. There are concerns that construction of the Three Gorges Dam will make the Chinese river dolphin go extinct.

Furthermore, it has been recently noted that dam construction might cause earthquakes, although data to back this up are still insufficient.

Table 1 Major Asian Dams

Name	Country	Year completed	Capacity (Mm³)
Baishan	China	1984	4,967
Dongjiang	China	1989	8,120
Gezhouba	China	1992	1,580
Liujiaxia	China	1968	5,700
Longyangsia	China	1988	24,700
Wujiangdu	China	1981	2,140
Yuecheng	China	1970	1,220
Balimela	India	1977	3,610
Bhakra	India	1963	9,621
Hirakud	India	1957	8,105
Idukki	India	1974	1,996
Pung Beas	India	1974	8,570
Ukai	India	1972	8,511
Mangla	Pakistan	1967	7,252
Tarbela	Pakistan	1976	13,690
Thac Ba	Vietnam	1971	3,600
Hoa Binh	Vietnam	1991	9,450
Tri An	Vietnam	1985	1,056

Source: United Nations, *Guidebook to Water Resources, Use and Management in Asia and the Pacific* (1995). I have chosen dams with capacities of 1 billion cubic meters or more.

Fig.1 Irrigated Area in Asia

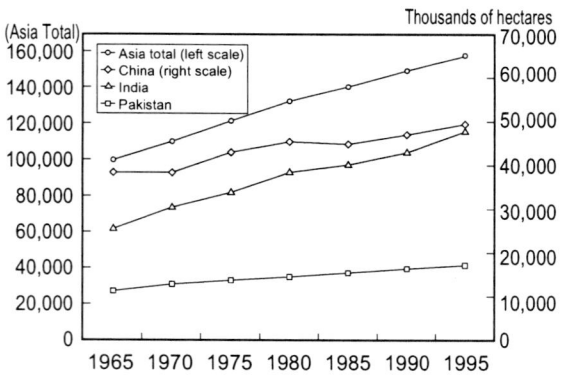

Source: Prepared from various years' editions of the FAO's *FAO Yearbook - Production.*

Fig.2 Japan's Use of Water for Industry

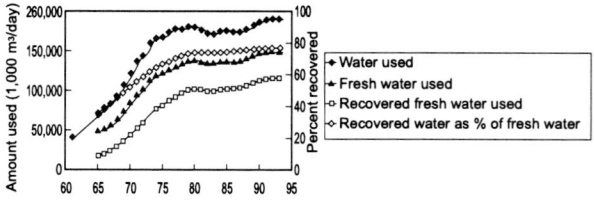

Source: Prepared from various years' editions of *Industrial Statistics Tables: Land and Water* (in Japanese).

Drawing water from rivers sometimes disrupts regional water cycles and has significant impacts on everything from river ecosystems to even the climate over broad regions. In Central Asia the rivers Amu Darya and Syr Darya, which flow into the Aral Sea, were diverted to irrigate cotton, and the reduction shrank the sea to about half its original area and one-fourth the amount of water. As a consequence, salt concentration increased and killed off nearly all aquatic life, the exposed lake bed turned into a vast desert, and the salt-carrying sandstorms arising there had damaging effects on the health of regional inhabitants. Some suspect that the sandstorms also carry pesticides used in growing cotton. Former fishing ports are now over 100 km away from the shore, so that in addition to ecosystem changes induced by higher salt concentration, the regional economy has suffered a crushing blow because it depended on the fishing industry.

In Japan water demand by the urban household and industrial sectors is 36% of total water demand, and it is thought that urbanization and industrialization in other Asian countries will likewise increase household and industrial water demand. Because water for household and industrial use necessitates higher quality than agricultural supplies, groundwater is often used to fill demand. Excessive use of groundwater causes ground subsidence, and recently there are reports that this is happening in Beijing, but there are no organized data on subsidence in Asia.

(YOSHIDA Hiroshi)

[17] Water Pollution Crisis

Currently agriculture consumes more water than any other use in Asia, and it is also a cause of water pollution. There are three main reasons for this. First is the eutrophication of lakes and the nitric acid contamination of groundwater by heavy fertilizer application. However, as rice paddy soil is reductive, it is said to resist nitric acid contamination by fertilizer. Second is pesticide contamination, and third is water contamination by animal wastes from large-scale stock farming operations. Animal wastes flow into lakes where they bring about eutrophication, and also cause nitric acid contamination of groundwater.

All around Asia lakes and tidal flats are being reclaimed. This results in the loss of areas with considerable water purification capabilities, and therefore worsens water quality. One example of reduction in lake area by reclamation is China's Lake Dongting (Table 1).

These agricultural burdens on bodies of water become serious with heavy applications of fertilizer and pesticides, or when stock farming is large in scale and intensive, i.e., when agriculture becomes a highly capitalistic enterprise. It is quite likely that the large plantations in Southeast Asia are causing eutrophication, as well as contamination from nitric acid and pesticides, but systematic data are still unavailable. The intensive farming of shrimp in Indonesia, Vietnam, and other countries for export to Japan brings about eutrophication from heavy feeding and water contamination from antibiotics.

Japan is the most industrialized and urbanized Asian country, and it has commensurate water pollution problems. Pollution by chemical substances from mining and industrial activities have been around since the Industrial Revolution came to Japan, but have been perceived as a major social issue since the 1960s. And because Japan's sewerage construction was delayed, untreated urban sewage flowed into and polluted the rivers. Industrialization and urbanization in other Asian countries are at stages more advanced than Japan was several decades ago, giving rise to concern that there is considerable water pollution by chemical substances (Table 2).

Mining is sometimes a cause of serious water pollution. In Japan, for example, the cadmium contamination that caused itai-itai disease in Toyama Prefecture and the arsenic contamination in Toroku, Miyazaki Prefecture resulted from mining. The Asia Arsenic Network got its start from the Toroku mine pollution opposition movement, and that network is inquiring into arsenic pollution in Thailand's

Nakhon Si Thammarat, China's Inner Mongolia Autonomous Region, India's West Bengal, and other places in Asia.

In Japan integrated circuit manufacturers disposed of the organic chlorine solvents trichloroethylene and tetrachloroethylene used in making LSIs by allowing them to seep into the ground, which contaminated groundwater in Fuchu City, Tokyo; Kimitsu City, Chiba; Taishi Town, Hyogo; and other places. There are concerns that such "hi-tech" pollution will spread throughout Asia.

Table 1 Change in Area of China's Lake Dongting

Year	Area (km^2)
1825	6,272
1896	5,400
1949	4,360
1954	3,915
1958	3,141
1973	3,915
1979	2,820
1983	2,691

Source: China Research Institute, ed., *China's Environmental Problems*, Shinhyoron.

Fig.1 Emission Surcharge Collections in South Korea

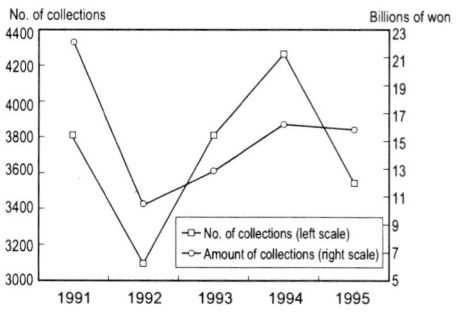

Source: South Korea's Environment Ministry, *Environment White Paper*, various years.

Table 2 Water Quality in Major Asian Rivers

River	BOD	Total P	N (nitric acid, nitrous acid)
Yangtze	0.8	0.05	1.15
Yellow	1.18	5.13	2.48
Han	1.3	0.02	-
Chao Phraya	0.5	-	-

Source: Kaya Yoichi, ed., *Environmental Databook 96/97.*

Another problem in Japan is contamination of drinking water when organic substances in source water combine with the chlorine added for purification and form trichloroethane and other carcinogens. Although such problems have yet to become serious in other parts of Asia, the likelihood of their occurrence makes it preferable to act soon to protect water sources.

Problems like those in Japan have arisen in South Korea, which is the next most industrialized and urbanized Asian country. To supply water to the industrial cities of Ansan and Shiheung an arm of the sea was closed off to create Shihwa Lake, but the delay in these cities' sewerage construction allowed untreated effluent to enter and pollute the lake, which was subsequently abandoned as a water source. South Korea's strict controls on the development of headwater areas have engendered inter-regional conflicts between water source and consumption areas. South Korea also uses a variety of economic means to deal with pollution (not all of which are firmly grounded in economic theory), representative of which is an emissions surcharge (Fig. 1).

In January 1997 a Russian tanker, the *Nakhodka*, sank in the Sea of Japan and released a large amount of heavy-grade heating oil. Tanker accidents like this one often pollute the ocean in Asia, calling for quick action to develop international measures to deal with them.

(YOSHIDA Hiroshi)

Table 3 Water Supply and Demand in Asia

		1	2	3	4	5	6	7
		China	Indonesia	Japan	S. Korea	Malaysia	Philippines	Thailand
A.	Average rainfall (mm)	600	2,000	1,750	1,200	2,500	2,250	1,550
B.	Water resources (km³)							
B.1	Renewable water resources (km³)	2,812	2,530	435	70	566	292	210
B.2	Above, per capita (m³)	2,360	13,230	3,522	1,919	31,851	4,690	3,778
C.	Water use							
C.1	Amount of water used (km³)	500	49	90	29	12	44	33
C.2	Above, per capita (m³)	435	338	731	658	653	717	591
C.3	Urban household use (%)	7%	9%	18%	22%	10%	12%	5%
C.4	Industrial use (%)	10%	2%	17%	11%	13%	4%	4%
C.5	Agricultural use (%)	83%	89%	65%	67%	77%	84%	91%
D.	Water supplies							
D.1	Groundwater (%)	18%	–	–	7%	25%	4%	–
D.2	Surface water (%)	72%	–	–	93%	75%	96%	–
E.	Irrigated area	493,680	4,597	27,800	13,350	3,400	15,800	48,000
F.	Tapwater connection rate							
F.1	Urban (%)	87%	35%	95%	100%	96%	93%	–
F.2	Rural (%)	68%	33%	–	76%	66%	72%	85%
G.	Fertilizer use							
G.1	Nitrogen (N) (thousands of tons)	19,791	1,700	588	475	336	403	740
G.2	Phosphoric acid (P_2O_5) (thousands of tons)	7,428	540	710	222	170	107	340
G.3	Potash (K_2O) (thousands of tons)	2,355	387	485	263	700	93	200
H.	Domestic animals							
H.1	Cattle (thousands)	100,859	11,595	4,916	3,075	689	2,021	7,593
H.2	Pigs (thousands)	424,680	8,720	10,250	6,100	3,282	8,941	4,507
H.3	Foul (millions)	2,798	640	315	81	100	74	80

Notes: B2 data for China, Indonesia, and the Philippines are 1992; South Korea, 1991; Malaysia, Thailand, and Japan, 1990. C data for China, Indonesia, Malaysia, the Philippines, and Thailand are for 1990; South Korea, 1993; and Japan, 1998. D data for South Korea are for 1993; others, 1990. F data for all countries are for 1990. Tapwater connection rate for Japan is national average.

Sources: A-D, F: United Nations, *Guidebook to Water Resources in Asia and the Pacific*, 1995.
E, H: FAO, *FAO Yearbook - Production*, 1995.
G: FAO, *FAO Yearbook - Fertilizer*, 1995.

[18] Wastes and Their Disposal

Accurate data on municipal solid waste and industrial waste for Asian countries are hard to come by. ESCAP has divided Asian countries into groups by income and used this to estimate how much MSW they generate (Table 1).[1] ESCAP's estimate puts the volume of MSW in the Asia-Pacific at about 700 million tons annually. Furthermore, some predictions say that continued economic growth will

----
[1]   United Nations Economic and Social Commission for Asia and the Pacific, *State of the Environment in Asia and the Pacific*, Chapter 14, 1995

approximately double that volume by 2005. Per capita waste generated in Australia, New Zealand, and Japan is far more than other Asia-Pacific countries. The higher a country's income level, the higher are the MSW disposal costs shouldered by the public sector, coming to $240-430 per ton, nearly five times the cost in low-income countries. MSW in countries that have industrialized contains high proportions of paper, plastic, and other packaging materials, while the proportion of food waste tends to diminish (Table 3).

Fig.1 Municipical Solid Wastes Generated by Type in Asian Countries

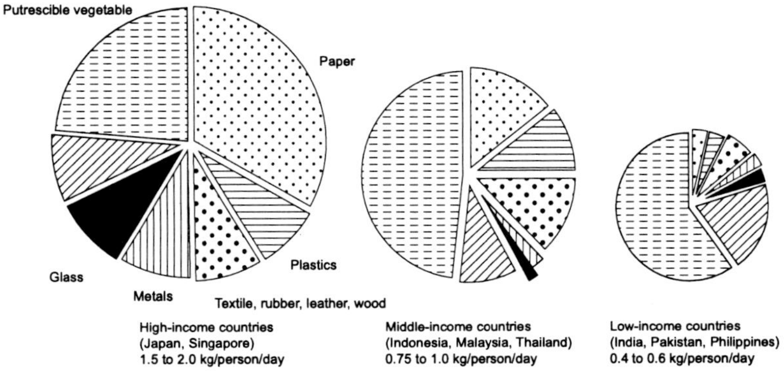

High-income countries
(Japan, Singapore)
1.5 to 2.0 kg/person/day

Middle-income countries
(Indonesia, Malaysia, Thailand)
0.75 to 1.0 kg/person/day

Low-income countries
(India, Pakistan, Philippines)
0.4 to 0.6 kg/person/day

Source: ESCAP(1995), Fig.14.1.

Table 1 Disposal Methods for Municipal Wastes in Selected Countries / Territories of the ESCAP Region

Country	Disposal method			
	Landfilling	Incineration	Composting	Others
Indonesia	80%	5%	10%	5%
Japan	22%	74%		4%
S. Korea	90%			10%
Malaysia	70%	5%	10%	15%
Philippines	85%		10%	5%
Thailand	80%	5%	10%	5%
Australia	96%	1%		3%
Bangladesh	95%			5%
Brunei	90%			10%
Hong Kong	65%	30%		5%
India	70%		20%	10%
Singapore	35%	65%		
Sri Lanka	90%			10%

Source: ESCAP(1995), Table 14.3.

Table 2 Reported and/or Estimated Annual Production of Hazardous Wastes in the Asia-Pacific Region

Country	1993 population (millions)	Reported/estimated hazardous wastes (thousands of tons)
China	1,176.9	50,000
Indonesia	186.4	5,000
Japan	124.7	82
S. Korea	43.8	269
Malaysia	18.3	377
Philippines	63.4	80-150
Thailand	58.1	882
Taiwan	20.9	3,000
Australia	17.3	109
India	873.0	39,000
Hong Kong	5.8	35
Singapore	3.1	28

Source: ESCAP (1995), Table 14.5.

It is estimated that between 30 and 50% of the MSW generated in the main areas of Asian cities is uncollected. The trash collectors and scavengers active in many Asian cities are believed to play a major waste disposal role in the informal sector, which has three important waste disposal functions. First, it is a way to make a living for the hundreds of thousands of urban poor excluded from society's formal sector. Second, it substantially reduces the amount of wastes that must be processed or disposed. And third, it recovers raw materials for production, thereby contributing to energy and resource conservation.

There are also large differences in MSW disposal methods (Table 1). Incineration rates in Japan and Singapore are high at around 70%, but landfilling is more common in other countries. In Asia as a whole the greatest concern is environmental and sanitation problems caused by inappropriate waste management, but other major problems include collection systems that cannot keep pace with increasing MSW amounts, inadequate processing, and too few landfill sites.

Currently it is impossible to obtain comprehensive, reliable data on industrial and hazardous wastes. According to tentative ESCAP estimates, about 100 millions tons of hazardous wastes are generated in the Asia-Pacific region each year, about nine-tenths of that amount in China and India (Table 2).

Estimates for recovery and recycling rates in the region are over 50% for both paper and glass, but little headway has been made in the recycling of hazardous materials such as batteries, waste oil, and plastic.

(YOSHIDA Fumikazu)

Table 3 Municipal Solid Wastes and Industrial Wastes Generated in Asian Countries

	1	2	3	4	5	6	7	8
	China	Indonesia	Japan	S. Korea	Malaysia	Philippines	Thailand	Taiwan
		Jakarta	Metropolitan Tokyo		Kuala Lumpur	Manila	Bangkok	
A. Municipal solid wastes [Year]	[1995]	[1994-95]	[1992]	[1994]			[1991]	
A1 Total amount generated (Thousands of tons/yr)	107,480	25,715 m³/day	50,199	21,213	730	1,360	1,856	8,490
A2 Per capita (kg/person/day)		1.1	1.1	1.35	1.29	0.50	0.77	1.1
A3 Percentages by weight (%)								
Paper		10	63	21	11.7	10.6	17	24
Glass		2	0	9 (briquettes)	2.5	2.3	6	9
Metals		2	1	5	6.4	3.6	6	8
Plastics		8	20		7.0	9.3	10	18
Fibers		2	3		1.3	4.8	3	4
Kitchen waste		74	6	31	63.7	31.8	16	23
Wood		1		4	6.5	10.8	4	8
A4 Recycling rate (%)			4	15				
A5 Incineration rate (%)			77	3	6.5	7-8		8
B. Industrial wastes [Year]	[1994]	[1992-93]	[1992]	[1994]				[1993]
B1 Total amount generated (Thousands of tons/yr)	617,040		403,481	32,460				43,760
B2 Percentages by type (%)	Mine tailings 30		Sludge 44	Slag 40				<Disposal methods>
	Coal cinders 19		Animal excreta 19	Demolition waste 14				Incineration 20
	Coal ash 17		Demolition waste 16	Combustion products 10				Chemical treatment 2
	Slag 12		Slag 8					Landfilling 38
	Metal slag 11							Recovery and reuse 42
B3 Recycling rate(%)	43		40					
C. Hazardous wastes (Thousands of tons/yr)	46,800			1,351		Discarded acids/alkalies 25 million m³ Discarded solutions 2000m³ Heavy metals 22,000 tons	933 Plating 50 Dyeing 14 Photography 16 Automobile batteries 12	1,200

Note: A3-2, 1991-92 figures; A3-3, burnable wastes, ratios by volume.

Sources: A-1: China Statistical Yearbook, 1996; B and C-1: China Statistical Yearbook, 1995.
A-2: BPS, Environmental Statistics of Indonesia 1995.
B and C-2: Ishii Akio, "A Collection of Data on Indonesia's Waste Management Industry and Associated Areas," in Cities and Wastes, 26-3, 1996 (in Japanese).
A-3: Ministry of Health and Welfare, Essentials of the Container and Packaging Recycling Law, 1995 (in Japanese).
B and C-3: Ministry of Health and Welfare, Waste Management in Japan, 1993 edition (in Japanese).
A-4 through C-4: 1996 Environment White Paper.
A-5 and 6: UNESCAP, State of the Environment in Asia and the Pacific, 1995.
B and C-6, B and C-7: IMO, Global Waste Survey, Final Report, 1995.
A-7: Japan International Cooperation Agency, The Study on Bangkok Solid Waste Management, 1991.
A-8 through C-8: Republic of China Environmental Protection Yearbook 83, 1994.

[19] Transfrontier Movements of Hazardous Wastes

Although the definition of "hazardous waste" differs from one country to the next, the OECD keeps statistics on the imports and exports of hazardous wastes by its member countries (Table 1), which show that for the total over the three years from 1989 to 1991 Germany exported the most, followed in order by Belgium and Holland, while Belgium was top in imports. The U.S. is listed as the sixth-place exporter, but Greenpeace documents indicate many more exports.

Greenpeace assembled official and unofficial data from a number of countries into a database on hazardous waste exports by OECD countries to non-OECD countries from 1989 to the spring of 1994, showing that Germany was the biggest exporter, followed by the U.S. and Britain. The receiving regions were, from largest to smallest, Eastern Eu-

rope and the former Soviet Union, the Asia-Pacific, and Latin America. Data on hazardous waste exports to the Asia-Pacific for 1990 to 1993 (Table 4) show that the U.S. is by far the largest exporter, followed by Canada and Britain. South Korea and India are the largest receivers. An overwhelmingly vast proportion of these wastes consists of metals such as lead, copper, and tin, with additional wastes including sludge, waste plastic, spent lead-acid batteries, leather waste, and discarded computers.

In line with a domestic law to implement the Basel Convention, Japan's hazardous waste import and export statistics for 1994 and 1995 have been released (Table 2). In the latter year there were 26 cases (representing 2,814 tons of wastes) in which exports were actually planned and documents submitted to the government, while for imports there

Table 1 Transfrontier Movements of OECD Countries' Hazardous Wastes (1989-1991, tons)

	Exports			Imports		
Year	1989	1990	1991	1989	1990	1991
Belgium	176,983	491,784	645,636	1,036,260	1,070,496	1,021,798
France	n.d.	10,552	21,126	n.d.	458,128	636,647
Germany	990,933	522,063	396,607	45,312	62,636	141,660
Holland	188,250	195,377	189,707	88,400	199,015	107,251
Britain	0	0	525	31,943	44,542	46,714
U.S.	118,927	118,416	108,466	n.d.	n.d.	n.d.

Source: OECD, *Transfrontier Movements of Hazardous Wastes, 1991 Statistics*, 1994, p. 9.

Table 2 Imports and Exports of Japan's Specified Hazardous Wastes (Controlled by the Basel Convention)

	Exports	
	FY1994	FY1995
Notifications to accepting country	7, for 10,320 tons (1)	8, for 1,791 tons (2)
Export approvals	3, for 7,580 tons	3, for 474 tons
Transfers of export documents	6, for 2,397 tons	26, for 2,814 tons

	Imports	
	FY1994	FY1995
Notifications from exporting country	8, for 1,329 tons (3)	21, for 4,179 tons (4)
Import approvals	7, for 1,216 tons	19, for 2,889 tons
Transfers of import documents	6, for 509 tons	47, for 1,163 tons

Notes: (1) Importing countries were France, the U.S., Belgium, Britain, Indonesia, and South Korea. All exports were for recovery of copper, tin, and other metals.
(2) Importing countries were France, Germany, Britain, the U.S., South Korea, and Malaysia. All exports were for recovery of copper, tin, and other metals.
(3) Exporting countries were the U.S., Holland, Malaysia, and the Philippines. All imports were for the recovery of cooper, zinc, silver, and other metals.
(4) Exporting countries/regions were Australia, Holland, the U.S., the Philippines, Hong Kong, South Korea, and Malaysia. Imports were for the purpose of recovering and reusing copper, silver, and other metals. One instance was for landfilling lead-containing sludge from plating effluent treatment.
Source: Environment Agency, Office of Marine Environment and Waste Management.

Table 3 Japanese Lead Waste Exports by Receiving Country/Region

(tons)

Year	1989	1990	1991	1992	1993	1994	1995	1996
China		933	3,204	4,918	539	15	51	18
Indonesia	992	8,832	13,319	11,222	3,379			649
S. Korea	3,465	2,464	1,033		34	150	804	5,721
Philippines	96	10	8	33	833	120		40
Thailand	1,413	720	4,330	4,104	2,500			
Taiwan	16,085	9,568	1,748	16			62	148
Hong Kong	1,902	2,376	1,602	245	205			6
Totals	23,971	25,528	25,261	20,614	8,162	304	1,195	8,018

Note: Other export receivers not listed.

Source: Prepared on the basis of *Japan Exports & Imports*.

were 47 cases (1,163 tons). The importers were France, Germany, Britain, the U.S., South Korea, and Malaysia, and all exports were for the stated purpose of recovering metals such as copper and tin. But these statistics are not necessarily an accurate representation of reality. For example, a considerable portion of the "lead waste" (code 7802) in *Japan Exports & Imports* is thought to be in spent lead-acid batteries, but this does not appear in Japan's Basel Convention statistics, which means such waste is not controlled by the domestic Basel legislation. "Copper waste" (7404), whose exports run to about 50,000 tons annually, are likewise un-

regulated. Statistics on Japan's exports of "lead waste" according to receiving country/region (Table 3) show that in 1989 and 1990 Taiwan was the biggest receiver, but pollution in that country triggered regulation that subsequently restricted imports, so Japan's exports to Indonesia increased in the 1990s. While enforcement of the Basel Convention effected a substantial overall reduction in exports in 1994, they have recently started increasing again, with exports to South Korea and other countries growing.

(YOSHIDA Fumikazu)

Table 4 Hazardous Waste Exports to Asian Countries/Regions, 1990-1994

(tons)	1 China	2 Indonesia	3 S. Korea	4 Malaysia	5 Philippines	6 Thailand	7 Taiwan	8 India	9 Hong Kong	10 Singapore	11 Bangladesh	12 Pakistan	13 Sri Lanka	Totals
A. By exporting country														
Australia	604	13,680		239	27,235	1,794	129	34,312	2,422	170	165			80,750
Canada	8,868	511	7,330		57		16,466	109,380	20,311			290		163,213
Britain	3,600	1,563	28,719	3,780	1,111	2,559	51,492	8,116	15,248	2,001	7	5,128	294	123,618
U.S.	220,665	20,490	3,016,496	325	35,932	93	198	1,779,282	42,899	71	3,146	6,885	425	5,126,907
Germany		620	19		50				2,338	240				3,267
Totals	233,737	36,864	3,052,564	4,344	64,385	4,446	68,285	1,931,090	83,218	2,482	3,318	12,303	719	5,497,755
B. Exports														
Ashes and slag		21				357	1,280	2,051						3,709
Iron		124						106,005				290		106,419
Hazardous substances	3,600		23,330	3,780					5,900	1,993		3,616		42,219
Nonferrous metals	97,392	14,776	3,016,681		27,309	379	5,470	1,829,644	9,348	469		7,406		5,008,874
Plastics	1,849	7,327	74	522	7,236	90	317	11,951	71,529	381	16	38	432	101,762
B*Others	65					1,737	45,000		42				294	47,138
C. Nonferrous metals breakdown														
Aluminum	265	87	4,480				17,323		420	100		609		23,284
Cadmium	176		220											396
Copper	6,600	103	4,670		50		4,770	5,980	2,582		3,130			27,885
Lead	2,228	14,428	1,034	42	29,873	1,884	123	1,479	172	8		36		51,307
Nickel	428		365											793
Tin											172	720		892
Zinc	6,517		18	720			4,005							10,540
Others	120,630	2,634						90						123,354

Notes: 9 and 10: Hong Kong and Singapore serve as transit points leading to China and other countries.
  B: According to loading documents.
  B*-1 and 9: Technojank.
  B*-6: Including radioactive wastes.
  B*-13: Hazardous wastes.
Source: Greenpeace, *The Waste Invasion of Asia*, 1994.

[20] Greenhouse Gas Emissions

The greenhouse gases (GHGs) that most contribute to global warming are CO_2, methane, CFCs and some of their substitutes, and N_2O. Especially because the contribution of CO_2 is said to be about half the total, and because its emissions are a reflection of economic activity, the gas is used as a global warming indicator.

The International Energy Agency (IEA) predicts that world CO_2 emissions in 2010 will be 48% higher than in

Table 1 Predicted World CO_2 Emissions

(Billions of tons CO_2)

	1971		1990		2000		2010	
OECD	9.1	61.3%	10.4	48.2%	11.8	47.1%	13.4	41.9%
Former Soviet Union, Central/Eastern Europe	3.3	22.0%	4.8	22.0%	3.9	15.5%	4.6	14.3%
China			2.4	11.1%	3.4	13.5%	5.0	15.7%
East Asia			1.0	4.6%	1.7	6.8%	2.6	8.2%
South Asia			0.7	3.2%	1.0	4.0%	1.7	5.3%
Others			2.4	11.1%	3.3	13.1%	4.7	14.7%
World	14.9	100.0%	21.6	100.0%	25.1	100.0%	32.0	100.0%

Source: Based on IEA, *World Energy Outlook 1994 edition.*

Fig.1 Per Capita CO_2 Emissions by Region (tons CO_2, 1992)

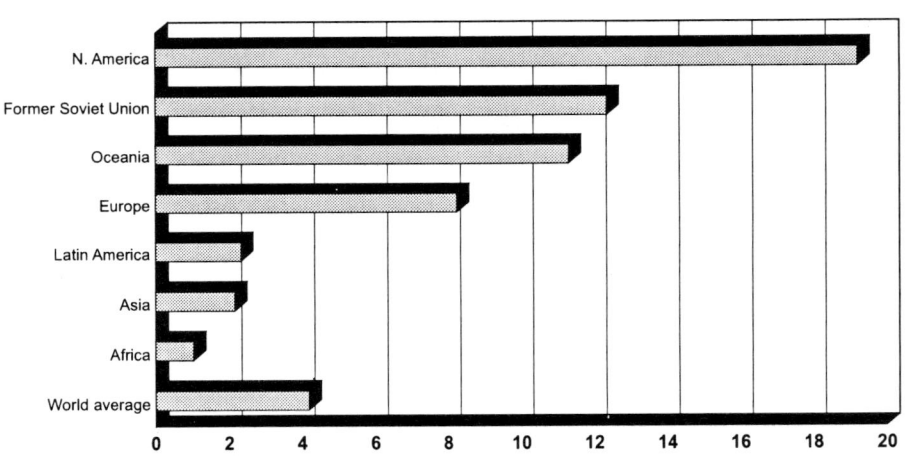

Source: WRI, *World Resources* 1996-97.

Table 2 Main Greenhouse Gases

	CO_2	CH_4	N_2O	CFC-11	HCFC-22
Pre-industrial concentrations	280ppmv	0.7ppmv	275ppbv	0	0
1994 concentrations	358ppmv	1.72ppmv	312ppbv	268pptv	110pptv
Recent annual concentration increases	1.5ppmv	10ppbv	0.75ppbv	0pptv	5pptv
	0.4%	0.6%	0.25%	0%	5%
Lifetime in atmosphere, years	50-200	12-17	120	50	12
Global warming potentials	1	25	320	4000	1700

Note: Concentrations of N_2O, CFC-11, and HCFC-22 are for 1992-93.
Source: IPCC, *Climate Change 1995.*

1990. In 1971 the OECD countries accounted for 61% of CO_2 emissions, but in 2010 the non-OECD countries' share will be 58%. Developing countries in particular will jump 2.6 times from 16.6% in 1971 to 43.8% in 2010, and 5.6 times in terms of total emissions (Table 1).

As East Asia and South Asia will emit 29.1% of the world's total CO_2 in 2010, trends in Asia will have a large impact on global warming (Table 1). The breakdown of CO_2 emissions for eight Asian countries indicates that emissions correspond to their energy consumption. For example, China and India burn high proportions of coal, while in the other countries oil predominates (Table 3).

In recent years the industrialized countries' share of total world CO_2 emissions has declined as the developing countries industrialize. But total emissions are rising fast, and the industrialized countries still bear substantial responsibility. In terms of per capita CO_2 emissions there is a clear North-South difference between industrialized and developing regions. Indeed there is a 19-fold difference between North America, which is highly industrialized, and Africa, where economic development is delayed (Fig. 1). An Asian example is the 22-fold difference between Japan and Vietnam.[1]

---

[1] WRI, *World Resources 1998-99.*

One cause of increased CO_2 emissions is deforestation (Table 2). Both tropical and temperate forests are subjected to heavy commercial logging in order to produce pulp and lumber, and in the developing world the opening of farmland by poor farmers is rapidly shrinking the area of forested land. Industrialized countries' economic activities and the widening North-South gap work to reduce CO_2 sinks and accelerate global warming.

Atmospheric concentrations of methane, CFCs, and N_2O are rising quickly. Especially the concentrations of CFC substitutes, which are powerful GHGs, are building at a fast pace (Table 2).

Methane emissions are high in countries with large populations like China and India, which is because the numbers of domestic animals, fields, and rice paddies are directly proportional to population. Methane is also emitted in large quantities when mining coal, and in recent years waste landfills have become another source.

CFCs and their substitutes are used as refrigerants, semiconductor cleaners, and blowing agents. There are hardly any emissions from Asian countries other than highly industrialized Japan (Table 3).

(UEZONO Masatake)

172

Table 3 Greenhouse Gas Emissions

		1 China	2 India	3 Japan	4 S. Korea	5 Malaysia	6 Philippines	7 Thailand	8 Taiwan
A.	CO_2 emissions from industrial processes (10,000 tons CO_2, 1992)								
	Industrial processes total	266,798	76,944	109,347	28,983	7,049	4,970	11,248	9,651
	Solid fuel combustion (coal, etc.)	208,801	55,190	31,779	9,261	621	516	1,583	3,613
	Liquid fuel (oil, etc.)	39,829	16,133	62,229	16,654	4,217	4,130	7,280	5,086
	Gas fuel (natural gas, etc.)	3,024	2,242	10,819	943	1,441	0	1,481	165
	Gas flaring	0	887	0	0	296	0	0	–
	Cement manufacturing	15,144	2,492	4,520	2,125	475	324	904	784
B.	Bunker fuel (10,000 tons CO_2, 1992)	0	269	3,237	0	0	0	0	–
C.	Land use (deforestation, etc.; 10,000 tons CO_2, 1992)	15,000	6,500	–	150	21,000	11,000	9,200	–
D.	Per capita CO_2 emissions (tons CO_2, 1987)	2.27	0.88	8.79	6.56	3.74	0.77	2.02	4.91
E.	CO_2 emissions, 1980-2010 (10,000 tons CO_2)								
	1980	147,659	31,510	99,661	13,557	2,931	3,664	3,664	8,061
	1992	255,381	69,250	119,446	31,877	7,694	4,763	10,626	14,290
	2000	370,430	–	122,378	46,899	12,824	7,694	18,320	20,518
	2010	555,096	–	124,210	67,418	23,083	15,022	32,976	28,946
F.	Makeup of CO_2 emissions by sector (%, 1987)								
	Energy conversion (electricity generation, oil refining, etc.)	24.3	44.3	39.7	16.5	30.2	43.7	31.6	34.3
	Industry (including cement)	49.7	32.6	27.7	36.9	32.3	19.2	20.5	40.3
	Transportation	4.1	15.8	18.3	15.0	32.8	17.4	40.2	15.6
	Residential, agricultural, commercial, etc.	21.9	7.3	14.3	31.6	4.7	19.7	7.7	9.8
G.	Methane (10,000 tons, 1992)								
	Total	4,700	3,300	390	140	96	190	550	–
	Solid wastes	89	260	190	31	10	33	10	–
	Coal mining	1,500	220	8	14	0	0	0	–
	Oil and gas production	26	83	4	–	55	–	13	–
	Wet rice agriculture	2,400	1,600	31	85	27	140	480	–
	Domestic animals	700	1,100	160	9	4	22	49	–
H.	CFCs (10,000 tons, 1991)	0.8	0.3	6.4	0.4	0.2	0.1	0.2	–

Notes: F: CO_2 emissions excluding biomass fuel. CO_2 emissions produced by electricity generation are counted under that sector, and electricity consumption in final consumption sectors is zero. Thus the energy conversion sector accounts for more emissions beccause this differs from calculations that count CO_2 emissions corresponding to electricity consumed in final consumption sectors, and that count conversion loss, self-consumption, and loss from power transmission and distribution in the generating sector.

A-F: Carbon dioxide emissions are all calculated as CO_2 tons (3.664 times the carbon contained).

H: CFCs are CFC-11 and 12.

Sources: A8 through D8 and F: Science and Technology Agency, National Institute of Science and Technology Policy, ed., *Asia's Energy Use and the Global Environment*, 1993 (in Japanese).

E: Ministry of International Trade and Industry, Natural Resources and Energy Agency, ed., *Asia Energy Vision*, 1995 (in Japanese).

H: WRI, *World Resources 1994-1995*.

All others: WRI, *World Resources, 1996-97*.

[21] Global Warming Impacts and the International Response

Predictions say that global warming will raise the sea level and inundate coastal areas, which would have a great impact on the world's major cities because they are concentrated in areas that are within a few meters above sea level. In Japan's case, a land area of 861 square km with 2 million people and 54 trillion yen in assets is below sea level at high tide, but if sea level rises 1 meter, those figures will all double. What is more, assuming a 1-meter sea level rise, the occurrence of storm surges or tsunamis would inundate the dwellings of 15.4 million people and 378 trillion yen in assets (Table 1). According to estimates, the cost of dealing with sea level rise would be over 10 trillion yen just to raise the height of seawalls.

In the last few years violent storms have caused a rapidly increasing amount of damage throughout the world. As the expression "billion-dollar storm" indicates, countries around the world have suffered heavy damage, which attained $20 billion in just the three years of 1990 to 1992 (Fig. 1). Many insurance companies went bankrupt because they were unable to pay the claims.

Frequent droughts and localized heavy rains, thought to be closely connected to global warming, are causing heavy damage to agriculture. In regions such as North America's Midwest region, known as the world's breadbasket, experts predict that global warming will impact heavily on food production and have serious consequences for Japan because of its high dependence on food imports.

There are forecasts that agriculture will be gravely affected in East Asia as well, and many researchers say that their countries will be directly affected by global warming (Table 2). While there are still many uncertainties, predictions of the impacts on China's crops by 2050 range from -78

Table 1 Land Area, Population and Assets Under Sea Level Now and After Sea Level Rise

Average sea level	At avg. sea level			At high tide			At times of storm surges or tsunamis		
	Area (km²)	Population (millions)	Assets (trillions of yen)	Area (km²)	Population (millions)	Assets (trillions of yen)	Area (km²)	Population (millions)	Assets (trillions of yen)
Present	364	1.02	34	861	2.00	54	6,268	11.74	288
0.3 m sea level rise	411	1.14	37	1,192	2.52	68	6,662	12.30	302
0.5 m sea level rise	521	1.40	44	1,412	2.86	77	7,583	13.58	333
1 m sea level rise	679	1.78	53	2,339	4.10	109	8,898	15.42	378

Source: Matsui Teijiro, et al., "Prediction of Inundation Effects on Japan's Coastal Areas in Conjunction with Sea Level Rise," in *Papers on Coastal Engineering*, vol. 39, 1992 (in Japanese).

Fig.1 Insurance Claim Payments for Storm Damage

(US$ billions)

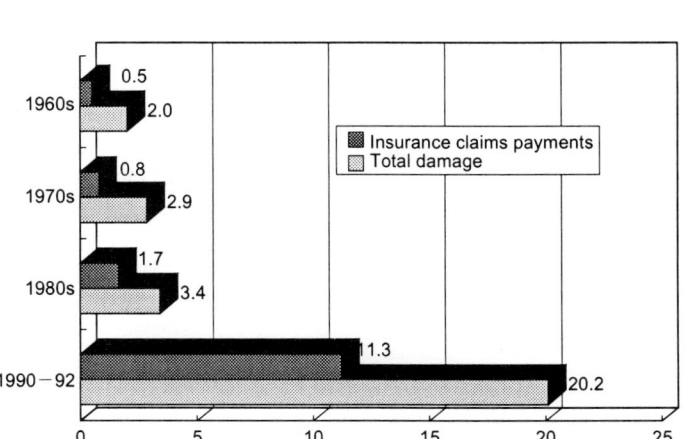

Note: Data from only storms causing at least $500 million in damages.
Source: IPCC, *Climate Change*, 1995.

Table2 Agricultural Yield Impact of Selected Climate Change Studies in East Asia

Country/Region	Yield Impact(%)	Direct CO_2 Effect	Author
Rice			
China	-6	Yes	Tao
China	-11 to -7	Yes	Zhang
China [1]	-78 to -6	No	Jin et al.
China [2]	-37 to +15	No	Jin et al.
China	-18 to -4	Yes	Matthews et al.
China	-21 to 0	No	Lin
Japan	+10	No	Sugihara
Japan	-11 to +12	No	Seino
Japan	-45 to +30	Yes	Horie
Japan	-28 to +10	Yes	Matthews et al.
Korea, Rep	-37 to +16	Yes	Yun
Korea, Rep	-40	Yes	Oh
Taiwan	+2 to +28	Yes	Matthews et al.
Wheat			
China	-8	Yes	Tao
China	-21 to +55	No	Lin
Japan	-41 to +8	Yes	Seino
Mongolia	-67 to -19	No	Lin
Russia [3]	-19 to +41	Yes	Menzhulin and Koval
Maize			
China	-4 to +1	Yes	Tao
China	-19 to +5	No	Lin
Japan	-31 to +51	Yes	Seino
Pasture			
Mongolia	-40 to +25	No	Bolortsetseg et al.

[1] This large negative percentage is a result of the modeled result at a single site in southwest China.
[2] For irrigated rice.
[3] Including European Russia.
Sources: IPCC, *The regional Impacts of Climate Change -An Assessment of Vulnerability-*, 1997, p. 371.

to +15% for rice, -21 to +55% for wheat, and -19 to +5% for maize. Another prediction is that in almost all areas of Japan the rice yield will decrease substantially with an atmospheric temperature rise of 4.0 to 4.5°C. Japan's non-rice crops, fruit, and pastureland will generally be adversely affected in the warm southwest, while in the north warming will in many cases speed photosynthesis and provide other benefits. But because global warming will stimulate further activity by weeds and insect pests, experts predict that this will diminish the combined advantageous effects so that the net overall effect on Japan's agriculture will be highly disadvantageous.

It is therefore very possible that global warming will bring about widespread damage. As a solution, countries have set goals to reduce their emissions of greenhouse gases in negotiations under the UN Framework Convention on Climate Change (Table 3). In 1997 FCCC parties adopted the Kyoto Protocol, under which Annex B countries (OECD countries and former Communist Bloc countries) are obligated to cut their collective greenhouse gas emissions 5.2 percent below 1990 levels by the period 2008 to 2012. As Japan is the only East Asian Annex B country, other East Asian countries are not obligated to make the same reduction, which means that for the time being East Asian countries will probably implement hardly any policies to combat global warming. In view of these circumstances Japan's role should be, in the international arena, to pursue active technical and financial aid, and, in the domestic arena, implement remedial global warming measures to ensure that its reduction obligation is met, and set a course toward cutting emissions that will be followed throughout the world.

(UEZONO Masatake)

Table 3 Abstract of the 1st National Communication of Industrialized Countries

Country	A Gas	B Standard year	C Object year	D National target	E 1990	E 2000 (estimate)	E Increase amount	E Increase rate(%)	E Result (1990=100)	F Stabilization target	F National target
1 Australia	GHGs	1988	2000	stabilization	288,965	332,799	43,834	15.1	-	no	no
	GHGs	1988	2005	-20%							
2 Austria	CO₂	1988	2005	-20%	59,200	65,800	6,600	11.1	99(1995)	prpbably	no
3 Canada	GHGs	1990	2000	stabilization	462,643	510,000	47,357	10.2	105(1994)	no	no
4 Denmark	CO₂	1990	2000	-5%	52,025	53,753	1,728	3.3	121(1994)	probably not	no
	CO₂	1988	2005	-20%							
5 Finland	CO₂	-	-	halting growth	53,900	70,200	16,300	30.2	108(1994)	no	no
6 France	CO₂	1990	2000	2t-C per capita	366,536	397,833	31,297	8.5	100(1993)	no	probably
7 Germany	CO₂	1990	2005	-25%	1,014,155	917,000	-97,155	-9.6	90(1993)	yes	no
	GHGs	1987	2005	-50%							
8 Greece	CO₂	1990	2000	+15%	82,100	94,500	12,400	15.1	-	no	no
9 Ireland	CO₂	1990	2000	+20%	30,719	36,988	6,269	20.4	107(1994)	no	yes
10 Italy	CO₂	1990	2000	stabilization	428,941	482,440	53,499	12.5	-	no	no
11 Japan	CO₂	1990	after 2000	stabilization per capita	1,155,000	1,200,000	45,000	3.9	107(1994)	no	no
	CO₂	1990	after 2000	stabilization of total emissions							
	CH₄	1990	1994	stabilization							
	other GHGs	1990	1994	stabilization							
12 Luxembourg	CO₂	1990	2005	-20%	11,343	7,556	-3,787	-33.3	-	yes	probably not
13 Netherlands	CO₂	1990	2000	-3%	167,600	167,600	0	0	105(1994)	no	no
	CH₄	1990	2000	-10%							
14 New Zealand	CO₂	1990	2000	stabilization	25,476	29,160-29,940	3,684-4,464	14.5-17.5	108(1994)	no	no
15 Norway	CO₂	1989	2000	stabilization	35,514	39,500	3,986	11.2	106(1994)	no	no
16 Spain	CO₂	1990	2000	+25%	227,322	276,523	49,201	21.6	-	no	very likely
17 Sweden	CO₂	1990	2000	stabilization	61,256	63,800	2,544	4.2	95(1994)	probably not chosen decrease	depends on chosen decrease target
	CO₂	1990	after 2000	decrease							
18 Switzerland	CO₂	1990	2000	stabilization	45,070	43,800	-1,270	-2.8	96(1994)	very probably chosen decrease	depends on chosen decrease target
	CO₂	1990	after 2000	gradual decrease							
	GHGs	-	-	decrease							
19 U.K.	GHGs	1990	2000	stabilization	577,012	586,720	9,708	1.7	96(1994)	maybe	maybe
20 U.S.A	GHGs	1990	2000	stabilization	4,957,022	5,163,136	206,114	4.2	103(1994)	no	no

Note: E: CO_2 emissions (CO_2 million kg). F: Estimate the targets.

Source: A-E: United Nations, *Framework Convention on Climate Change*, FCCC/CP/1996/12/Add.1, 28 June 1996; United Nations, *Framework Convention on Climate Change*, FCCC/CP/1996/12/Add.2, 2 July 1996; United Nations, *Climate Change Bulletin*, Issue 11, 2nd Quarter 1996.
F: Climate Network Europe and United States Climate Action Network, *Independent NGO Evaluations of National Plans for Climate Change Mitigation -OECD Countries*, 4th Review, June 1996.

[22] Environmental Laws and Legislation

Since the 1972 UN Conference on the Human Environment, the countries and regions of East and Southeast Asia have been pursuing environmental legislation. But because in the initial stages it was often the case that legal institutions were inadequate and ineffective, courts were not very effective in remedying environmental problems.

For these reasons from the end of the 1980s through the first half of the 1990s Asian governments embarked on the sweeping reform and strengthening of their legal institutions due to circumstances including (1) more and worse pollution, (2) a rising environmental consciousness both at home and abroad, and (3) the broad international diffusion of the sustainable development concept.

Of course the many differences in legal systems make comparison difficult. To begin with, there is a wide range of variation in the concept of "the environment," which translates into big differences among countries and regions in environmental law. China, for example, includes cities within the environment. There are also differences in the forms that laws take: Some countries think legislation is important, while some make frequent use of government ordinances. In countries that are federations, regulations sometimes differ from one region to another.

Yet, the current environmental law systems of East and Southeast Asia do share some characteristics.

First, they have basic environmental laws that serve as the basis for other laws meant to control pollution and protect the natural environment. In many countries, measures for conservation and the idea of sustainable development are incorporated into national economic and development plans, or into long-term planning. For example, in China the integrated achievement of economic, social, and environmental results is a fundamental aim under national policy.

Not a few countries include environment-related provisions in their constitutions, such as provisions for environmental rights (as in the Philippines and South Korea) and the state's obligation to take the environment into consideration (China).

Second, these countries and regions have established ministry- (as in South Korea and Malaysia) or agency- (Japan) level organizations, or environmental administrative organizations directly answerable to the president (like Indonesia's Environmental Impact Management Agency), thereby facilitating the transition from mere coordinating organizations to those with stronger authority.

Third, in conjunction with the broadening acceptance of environmental management-like thinking, there is a trend toward diversification in legal means, from regulatory mechanisms using permits and penalties, to economic methods (such as surcharges, preferential tax treatment, and commendations for outstanding companies). Some countries additionally endeavor to heighten the efficacy of traditional regulatory devices, such as by raising the amounts of fines, or providing for imprisonment (as in Thailand). Chinese authorities give polluting enterprises a time limit to make improvements, and require them to relocate, shut down, or take other action in case of non-compliance. And in the area of environmental and emission standards, there are now countries like Malaysia that establish industrial effluent standards tougher than those in Japan.

Fourth, many countries have preceded Japan in the institution of comprehensive environmental impact assessment laws that cover a variety of enterprises. In China, for example, failure to undergo an EIA screening stops construction approval, expropriation of land, loans from financial institutions, and the like. An enterprise that starts construction without a screening is required to stop construction, undergo screening, and pay a fine.

Fifth, in taking remedial action against pollution and in helping victims, not a few countries have adopted the polluter pays principle, and sometimes absolute liability (China and South Korea, for example) or strict liability (Indonesia, for example).

In addition to these shared attributes, the unique characteristics of each country's legal system also deserve examination.

Malaysia, for instance, as a rule bans the incineration of industrial wastes, allowing only final disposal at permitted sites. An increasing number of countries allow the participation of citizens and environmental NGOs in various ways, such as Indonesia, which under the law guarantees citizen participation in EIAs (i.e., participation in deliberations of the AMDAL[1] Commission in their localities) and the Philippines (participation in public hearings), as well as Thailand, which instituted a registration system for environmental NGOs, and passed a law under which the registered NGOs are supplied with information and given financing from the Environmental Fund (under the finance minister), as well as privileges including the right to recommend appointments to the National Environmental Board (NEB),

---

1 *Analisis Mengenai Dampak Lingkungan*, or EIA in Indonesian.

and Malaysia, which allows participation by environmental NGO representatives in an advisory body, the Environmental Quality Council (EQC).

There are also countries such as South Korea and Taiwan that deal with pollution disputes through alternative dispute resolution bodies similar to Japan's Environmental Dispute Coordination Commission, and those that, like China, allow class action-like lawsuits.

Among South Asian countries, which are not covered in this volume, there are countries with advanced legislation like India's Public Liability Insurance Act, whose features include compulsory insurance and an absolute liability system.

However, there is no telling to what extent these reforms by Asian governments will close the gap between written laws and actual situations. With EIAs, for instance, the Philippines and Indonesia have indeed carried out several thousand assessments each, but it is sometimes observed that their systems are not functioning effectively owing to factors such as inadequacies in the technical and specialized knowledge of assessment personnel, and citizen participation in name only. Despite tightened controls and strict environmental standards in Thailand, untreated industrial effluent apparently continues to run directly into rivers, and Malaysia has a serious shortage of industrial waste disposal sites with operating permits.

Justice in the courts of East and Southeast Asia for pollution victims is achieved primarily by seeking compensation for damages, but not very many lawsuits are filed. And despite the possibility of civil lawsuits seeking injunctions and administrative environmental lawsuits, in some countries not a single such lawsuit has been filed. More efforts are needed to ensure the efficacy of environmental law, including improvements in the capabilities of administrative personnel, plus environmental education for citizens.

(SAKUMOTO Naoyuki, OKUBO Noriko)

Table 1 Environmental Laws in Asia

	1 China	2 Indonesia	3 Japan	4 S. Korea	5 Malaysia	6 Philippines	7 Thailand	8 Taiwan
A. Managing administrative agencies	State Environmental Protection Administration	State Ministry of Environment Environment Impact Management Agency	Environment Agency	Environment Ministry	Ministry of Science, Technology and Environment	Department of Environment and Natural Resources, Environment Management Bureau	Ministry of Science, Technology and Environment National Environmental Board	Environmental Protection Administration
B. Basic environmental laws	Environmental Protection Law	Law Concerning Environmental Management Act No.23 of 1997 concerning the Management of the Living Environment	Basic Environment Law	Basic Environmental Policy Act	Environmental Quality Act	Environmental Policy (Presidential Decree No. 1151) Environment Code (Presidential Decree No. 1152)	Enhancement and Conservation of National Environmental Quality Act	Basic Environmental Conservation Law
C. Environmental impact assessment laws	Regulation on Construction Project Environmental Protection	Government Regulation No. 51 of 1993 on Environmental Impact Analysis	Environmental Impact Assessment Law	Environmental Impact Assessment Act	Environmental Quality (Prescribed Activities) (Environmental Impact Assessment) Order	Presidential Decree No. 1586, Establishing an Environmental Impact Statement System Including Other Environmental Management (Related Research and Other Purposes)	Notice of the Ministry of Science, Technology and Environment concerning Prescription of Types and Capacity of Projects or Activities of Government Agencies, State Enterprises or Private Sector Requiring Creation of Reports on Environmental Impacts (1996)	Environmental Impact Assessment Act
D. Water laws	Law on the Prevention and Control of Water Pollution	Government Regulation No. 20 of 1990 on Water Pollution Control	Water Pollution Control Law	Water Quality Preservation Act	Environmental Quality (Sewage and Industrial Effluents) Regulations	Presidential Decree No. 1067, Water Code (Revising and Consolidating the Laws Governing the Ownership, Appropriation, Utilization, Exploitation, Development, Conservation and Protection of Water Resources)	Factory Act Groundwater Act	Water Pollution Control Act
E. Air laws	Law on the Prevention and Control of Air Pollution		Air Pollution Control Law	Air Quality Preservation Act	Environmental Quality (Clean Air) Regulations	Presidential Decree No. 1181, (Providing for) Prevention, Control and Abatement of Air Pollution from Motor Vehicles and for Other Purposes (Anti Smoke-Belching Law of 1977)		Air Pollution Control Act

Table 1 Continued.

F. Noise/vibration laws		Noise Regulation Law / Vibration Regulation Law	Noise and Vibration Control Act	Environmental Quality (Motor Vehicle Noise) Regulations	Noise Control Act	
G. Waste laws	Law on the Prevention and Control of Solid Wastes	Government Regulation No. 19 of 1994 on Management of Hazardous and Toxic Wastes, as amended by Government Regulation No. 12 of 1995	Wastes Disposal and Public Cleansing Law	Waste Management Act	Environmental Quality (Scheduled Wastes) Regulations	Hazardous Substance Waste Disposal Act
H. Nature protection laws	Law on Wild Animal Protection / Grassland Law / Forestry Law	Act No. 5 of 1990 on Conservation of Natural Resources and Ecosystems, dated 10 August 1990 / Presidental Decree No.32 of 1990 on Management of Protected Areas / Decree of the Minister for Forestry and Plantations No. 12/Kpts-II/1987 on Wild Animal Protection, Government Regulation No. 27 of 1991 on Wetlands	Natural Parks Law / Nature Conservation Law	Natural Environment Preservation Act / Wetlands Conservation Act	National Park Act / National Parks Act / Protection of Wild Life Act	Wildlife Conservation (and Protection) Act / National Preserve Forest Act
I. Environmental dispute settlement laws		Environmental Pollution Disputes Settlement Act	Environmental Dispute Settlement Act			Pollution Dispute Management Law

Note: Indonesia and Malaysia have plans to amend a number of laws.

Sources: Prepared on the basis of the following sources with information on amendments added.

Nomura Yoshihiro and Sakumoto Naoyuki, eds., *Environmental Laws in Developing Countries: East Asia*, Institute of Developing Economies, 1993 (in Japanese).

Nomura Yoshihiro and Sakumoto Naoyuki, eds., *Environmental Laws in Developing Countries: Southeast and South Asia*, Institute of Developing Economies, 1994 (in Japanese).

Editorial Board of China Environmental Yearbook ed., *China Environmental Yearbook 1995*, China Environmental Yearbook, Inc., 1996.

C. Cory, L. Kurukulasuriya, W. Kiatsinsap, V. Sirisumpam, and P. Sukonthapan, ed., *Southeast Asia Handbook of Selected National Environmental Laws*, Mekong Region Law Center (MRLC), 1998.

URLs on environmental laws and administration:

China, http://www.qis.net/chinalaw/

Indonesia, http://www.geocities.com/RainForest/Vines/9160/

Japan, http://www.eic.or.jp/eanet/index-e.html

South Korea, http://www.me.go.kr/english/index.html

Philippines, http://www.denr.gov.ph/

Taiwan, http://www.epa.gov.tw/english/

Singapore (Asia-Pacific Centre for Environmental Law, Faculty of Law,

National University of Singapore), http://sunsite.nus.edu.sg/apcel/dbase/asean.html

[23] Membership in Environmental Conventions

Like many other countries in the world, East and Southeast Asian countries find increasing importance in international environmental law as a means of coping with global environmental problems (such a global warming), pollution export from other countries (like the transfrontier shipment of hazardous wastes), environmental problems that affect surrounding countries (acid rain, for example), and the like. Let us discuss how such conventions relate to Asia.

Many Asian countries are parties to major international conventions for protecting the oceans (MARPOL and the International Convention on Civil Liability for Oil Pollution Damage), the atmosphere (Vienna Convention and Montreal Protocol), and climate systems (Framework Convention on Climate Change or FCCC and Convention to Combat Desertification), and to control wastes (Basel Convention, London Dumping Convention) (Table 1).

Asian countries have significant interest in a number of conventions for protecting nature, wildlife, and cultural heritages such as the Biodiversity Convention and CITES. In this area there are also bilateral treaties between Asian countries, like the treaty for migratory bird protection between Japan and China. In this connection the Ramsar Convention is also of great importance to Asia because "wetlands" include rivers, rice paddies, and the like, thereby linking this convention to much of the development underway in developing countries. Asia also has many cultural assets that deserve protection by the World Heritage Convention, and a number of sites have been registered. Of great importance in conserving tropical rainforests is the 1994 International Tropical Timber Agreement, a commercial agreement with purposes that include limiting trade to timber produced in sustainably managed forests by 2000.

Coping with transboundary pollution will henceforth increase the necessity for regional conventions. An existing example is a four-party convention on Mekong basin development among Cambodia, Laos, Thailand, and Vietnam called the Agreement on Cooperation for the Sustainable Development of the Mekong River Basin. To deal with acid rain in this region, governments are now setting up the Acid Deposition Monitoring Network in East Asia.

Although the provisions of conventions do not apply to non-parties, nearly all countries in East and Southeast Asia are parties to, or have at least signed, the international conventions mentioned above. Nevertheless, the constant North-

Table 1 Membership in Environmental Conventions

	1 China	2 Indonesia	3 Japan	4 S. Korea	5 Malaysia	6 Philippines	7 Thailand	8 India
A. MARPOL Convention	○	○	○	○	×	×	×	○
B. Vienna Convention	○	○	○	○	○	○	○	○
C. Montreal Protocol	○	○	○	○	○	○	○	○
D. Framework Convention on Climate Change	○	○	○	○	○	○	○	○
E. Convention to Combat Desertification	○	×	×	×	○	○	×	○
F. Basel Convention	○	○	○	○	×	×	×	○
G. Biodiversity Convention	○	○	○	○	○	○	×	○
H. CITES	○	○	○	○	○	○	○	○
I. Ramsar Convention	○	○	○	○	○	○	×	○
J. World Heritage Convention	○	○	○	○	○	○	○	○
K. International Tropical Timber Agreement	○	○	○	○	○	○	○	○

Note: Data for D through K were prepared from each convention's web site (as of June-October 1997).

Sources: A-C: H.O. Bergesen/M. Norderhang/G. Parmann ed., *Green Globe Yearbook*, Oxford, 1994.
   D: UNEP, http://www.unfccc.de/
   E: UNEP, http://www.unccd.ch/
   F: UNEP, http://www.unep.ch/basel/index.html
   G: UNEP, http://www.biodiv.org/
   H: UNEP, http://www.unep.ch/cites.html
   I: IUCN, http://www.iucn.org/themes/ramsar/
   J: UNESCO, http://www.unesco.org/whc/
   K: ITTO, http://www.itto.or.jp/

Table 2 Registered World Heritage Sites in Major East Asian Countries (As of February 20, 1997)

China		Indonesia / others	
1987	The Great Wall	1991	Prambanan Temple compound
1987	Mount Taishan	1996	Sangiran Early Man Site

China
1987	The Great Wall
1987	Mount Taishan
1987	Imperial Palace of the Ming and Qing Dynasties
1987	Mogao Caves
1987	Mausoleum of the First Qin Emperor
1987	Peking Man Site at Zhoukoudian
1990	Mount Huangshan
1992	Jiuzhaigou Valley Scenic and Hisoric Interest Area
1992	Huanglong Scenic and Historic Interest Area
1992	Wulingyuan Scenic and Historic Interest Area
1994	The Mountain Resort and its Outlying Temples, Chengde
1994	Temple of Confucius, Cemetory of Confucius, and Kong Family Mansion in Qufu
1994	Ancient Building Complex in the Wudang Mountains
1994	The Potala Palace, Lhasa
1996	Lushan National Park
1996	Mt. Emei and Leshan Giant Buddha
1997	Old Town of Lijiang
1997	Ancient City of Ping Yao
1997	Classical Gardens of Suzhou
1998	Summer Palace, an Imperial Garden in Beijing
1998	Temple of Heaven, an Imperial Sacrificial Altar in Beijing

Indonesia
1991	Komodo National Park
1991	Ujung Kulon National Park
1991	Borobudur Temple compound
1991	Prambanan Temple compound
1996	Sangiran Early Man Site

Japan
1993	Himeji-jo
1993	Buddhist Mounuments in the Horyuji Area
1993	Yakushima
1993	Shirakami-Sanchi
1994	Historic Monuments of Ancient Kyoto (Kyoto, Uji and Otsu Cities)
1995	Historic Village of Shirakawa-go and Gokayama
1996	Hiroshima Peace Memorial (Genbaku Dome)
1996	Itsukushima Shinto Shrine
1998	Historic Monuments of Ancient Nara

South Korea
1995	Sokkuram Grotto and Pulguksa Temple
1995	Haiensa Temple Chaggyong P'ango, the Depositories for the Tripitaka Koreana Woodblocks
1995	The Chongmyo Shrine

The Philippines
1993	Baroque Churches of the Philippines
1993	Tubbataha Reef Marine Park
1995	Rice Terraces of the Philippines Cordilleras

Thailand
1991	Thungyai-Huai Kha Khaeng Wildlife Sanctuaries
1991	Historic Town of Sukhothai and Associated Historic Towns
1991	Historic City of Ayutthaya and Associated Historic Towns
1992	Ban Chiang Archaeological Site

Sources: UNESCO website (http://www.unesco.org/whc/heritage.htm and http://www.unesco.or.jp/sekaiisan/wolist.htm)

South antagonism in convention negotiations over issues such as who takes responsibility, provisions of funds, and conditions for technology transfer make it difficult to arrive at new agreements. Recently, therefore, negotiating parties facilitate agreements by using framework conventions (examples being the Bonn Convention, Vienna Convention, FCCC, and Biodiversity Convention), and then decide the specifics in protocols after the conventions have been concluded. There is also an increase in the use of "soft law," which involves devices like recommendations and declarations instead of legally binding conventions.

In not a few instances developing countries are accorded preferential treatment of some kind to relax North-South tensions. Examples of these are the FCCC's "common but differentiated responsibilities," the reduced obligations of developing nations as under the Montreal protocol, the provision of "new and additional financial resources" and "fair and most favorable terms" for technology transfer under the Biodiversity Convention.

But even when international agreements are reached, getting countries to comply is another matter. For example, many parties to CITES have yet to pass domestic legislation as called for by the convention, and violations occur continually. Hereafter in addition to encouraging international agreements and the signing of conventions, governments need to pass domestic legislation to put teeth into international agreements.

(SAKUMOTO Naoyuki, OKUBO Noriko)

Translator's Afterword

Welcome to Asia. For those of you new to this part of the world, we hope this book — the first in a series — will be a good introduction to the environmentally related problems and potentials of this region. For those already familiar with Asia, it is our hope that this volume will provide new insights into a facet of Asia that gets relatively little press coverage in the West.

This book is not a word-for-word translation of the original. In places I have cut and spliced, edited somewhat, and rearranged material slightly in an effort to make it more presentable in English instead of sticking slavishly to the Japanese original. Nevertheless, because I did this translation single-handedly and began well after the Japanese version was already on sale, time constraints did not afford the luxury of more editing and rewriting. We hope to remedy this problem in future editions.

Moreover, this first edition represents the first such effort for JEC as well as for me. In that respect it has been a real learning process, and we all hope that future editions will benefit from both the successes and failures of this one.

What we really need to make this series better in English is feedback from you, the reader. For future editions we need to know specifically what you want to read, and how you would like it presented. That being impossible for this first edition, all changes are entirely the product of my arbitrary, on-the-spot judgments made during the translation process. Constructive feedback is welcomed and should be directed to the email addresses given below.

The original contains many more notes and references than given in the footnotes. Except for the figures and tables in Section III, I have omitted nearly all references to sources in Asian languages under the assumption that they would be linguistically inaccessible to Western researchers. Those references could, however, be included in future editions if there is sufficient demand.

Finally, I want to thank the writers for their cooperation in looking over the draft translations of their respective sections. Despite their busy schedules, they took the time to check spellings of proper names and respond to my numerous, annoying questions about facts and content. A few were even gracious enough to rewrite short sections and add more recent information especially for this English-language version. Nevertheless, I should like to emphasize that any errors are my responsibility alone. Special thanks go to Oshima Kenichi for creating the charts and tables in an unbelievably short time. If any errors remain it is certainly because we rushed him so.

Rick Davis
Ashigawa, Japan, September 1999

Email contacts:
Rick Davis: rdavis@yin.or.jp
TERANISHI Shun'ichi: ce00024@srv.cc.hit-u.ac.jp

Index